THE BIOLOGIST'S MISTRESS

Rethinking Self-Organization in Art, Literature, and Nature

THE BIOLOGIST'S MISTRESS

Rethinking Self-Organization in Art, Literature, and Nature

Victoria N. Alexander

3810 N 188th Ave
Litchfield Park, AZ 85340

The Biologist's Mistress:
Rethinking Self-Organization in Art, Literature, and Nature
Written by: Victoria N. Alexander

Library of Congress Control Number: 2011931822

ISBN: 978-0-9842165-5-0

Printed in the United States of America

To Nathan Roy

ABOUT THE AUTHOR

Victoria N. Alexander is co-founder of the Dactyl Foundation in New York City, an arts organization dedicated to bringing the sciences into the arts and the arts into science. As a visiting researcher at the Santa Fe Institute, she studied evolutionary theory and complexity science and received a PhD in English from City University New York. She is also a novelist whose critically acclaimed titles include *Smoking Hopes, Naked Singularity*, and *Trixie.*

CONTENTS

PART II
HISTORY: SELECTED FIGURES

PART III
APPLICATION:
THE INFLUENCE OF TELEOLOGY ON STORIES

PREFACE

This book is about *teleology*, the study of the purposes in nature that make life (seem like) a meaningful work of art. Looking at a number of different teleologies in light of recent advances in the sciences of complexity and non-Darwinian evolutionary theory, I show teleology's long, little appreciated relationship to what I call *the emergent ordering tendencies of chance*. Along the way, I hope to rescue teleology from theology, which has kept it mired in confusion for centuries, and reconnect it to artistic practice.

Teleology is such a common thread throughout human culture that many may not be aware this type of research has a name. *Oh that!* some will say when they are done with this book, *I know what that is; I just didn't know it was called "teleology."* This branch of philosophy is so immensely influential that it touches on a little bit of everything, and so sometimes it's almost harder to say what it doesn't cover than to say what it does. Virtually every discipline has been engaged with it at one point: religion, philosophy, evolutionary theory, biology, ecology, psychology, cosmology, physics, chemistry, literary theory and so on, maybe even dentistry, I wouldn't be surprised. No sane person can be expected to be familiar with all or even most or even much of what has been written on teleology. So in an effort to make this work accessible and interesting to a wide audience, I have not included thorough histories or summaries of current research on teleology and intentionality (please consult the bibliography for further reading). Instead, I have chosen to include only brief references to only those scholars I have found especially useful in shaping my thought.

In my narrative, I tend to favor medieval philosophers over today's analytic philosophers, pre-Darwinian biology over 20th century developmental systems biology, biosemiotics over teleosemantics, the complexity sciences over general systems theory, neuroscience over psychology, pragmatism over deconstruction, and fiction over physics. No doubt I omit many important voices. I did not want to rehearse various theses only to say how I disagree. Let those who know the work make those comparisons themselves.

In some cases, you may feel that my omission indicates an appalling ignorance rather than an informed preference, and you might be right. It may very well be that I haven't read your key works and authors. But the wonderful thing about teleology today is that many specialized fields—fields that haven't interacted for decades—are simultaneously converging upon similar ideas about emergence, complexity, selfhood, and purpose. This may count as evidence that we are discovering something true, however much we may differ over finer details or reject each other's specific rhetorical choices.

I thank my friend, Michelle Fletcher, for her part in the stimulating conversations that helped begin this research. I thank the Dactyl Foundation, the Jewish Foundation for the Education of Women, the Santa Fe Institute, and the Art and Science Laboratory for supporting the original research for this book. I especially thank my four graduate school advisors who didn't balk when I chose such an unpromising topic as "teleology" to begin my career as an academic. I acknowledge Jim Crutchfield for his patience and willingness to help me understand the strange new world of deterministic chaos, computational mechanics, and complexity. I acknowledge Eve Kosofsky-Sedgwick's kind encouragement. And, above all, I am indebted to Angus Fletcher and Joan Richardson, whose own work confirms my belief that the study of literature and art is indispensable to science.

PART I

THEORY: WHAT IS TELEOLOGY?

Chapter 1

TELEOLOGY AFTER POSTMODERNISM

> *...nature is already, in its forms and tendencies, describing its own design. Let us interrogate the great apparition, that shines so peacefully around us. Let us inquire, to what end is nature?*
>
> –R. W. Emerson

Teleology is the study of the purposes of action, development and existence. Its practitioners believe *nature is purposeful.* An ancient and enduring form of inquiry that has been out-of-fashion among educated people for centuries, teleology's slow, steady decline as a scientific discipline began in the 17th century with the birth of modern empiricism and continued to plummet apace the rise of the Enlightenment, Darwinism, and quantum mechanics. Nature is not purposeful, it was said, and those who continued to think it was were primarily spiritualists, artists, or madmen, who credited the guidance of gods, muses, or fate.

Biologists—whose subject compels them to deal with questions about, for example, what organs are *for*—must constantly remind themselves that officially functionality is just a side effect of predictable material causal processes. As J. B. S. Haldane is said to have claimed, teleology is like a mistress to a biologist: he may not be able to live without her but he's unwilling to be seen with her in public. The serious and sensible scientists resolutely resist teleology and her meretricious allure. And so despite biology's occasional flirtation, in general science measures its progress in terms of the distance it has put between itself and teleology.

I call myself a teleologist, and in doing so risk a certain amount of professional shame and disrepute. When I was deciding on a career as a literary theorist and philosopher of science and entering graduate school, if any of my peers talked of teleology at all, it was only to say how passé or stilted "teleological" narratives were, on par with calendar art or sermons. People assumed I was religious or a Conservative or simply had bad taste.

Many of those who would receive me—some of whom also called themselves teleologists—were very unwelcome bedfellows. They talked of Truth, Beauty and Goodness and asked me to supply them with a Theory that would defend their particular ideas of T, B & G. Teleology concerns form and function, which is not the same thing, quite, as beauty and goodness, certainly not the same thing as Beauty and Goodness. So I was ultimately unwelcome in that group too.

What I do share with all teleologists, authentic or so-called, is a deeply felt folk-sense of purposefulness in nature. It is clear to me that many processes and patterns in nature can't be fully explained by Newton's laws or by Darwin's mechanism of natural selection. These are processes that are organized in ways that spontaneously create, sustain and further that organization. Although I believe that mechanistic reductionism is inadequate to describe these processes, I don't believe that purposeful events and actions require guidance from the outside—from divine plans or engineering deities. Nature's purposeful processes are *self*-organizing and *inherently* adaptive, which is the essence of what it is to be teleological.

A few examples: 1. Flocks of birds fly in formation and change direction simultaneously, even though there is no one leader in the flock nor any kind of instantaneous communication among the entire flock. 2. Many species appear to have been formed according to the same general ground plan: for example, many animals' major organs have relatively similar distribution schemes, even though they *do not* share a common ancestor. 3. When food resources are scarce, free-roaming slime-mold cells (if you do any amount of reading in science you know that slime mold and fruit flies have a kind of celebrity status) will emit a chemical signal that attracts other cells. They aggregate and finally pile up to form stalks that eventually release spores *in order to* continue the species-individual.

Each of these examples involves a process that appears to be guided by a plan or that anticipates the future. Yet, argue scientists, the individual birds, separate species, or free-roaming cells are not intentionally acting as an organized group, variations on a theme, or altruistic stalk builders. Such phenomena, it has been argued for four centuries of science, merely *appear* purposeful.

Instead, I wondered if these examples from nature can help us re-imagine what purposeful behavior actually is, in ourselves as well as in nature. I decided to pursue that thought. Against the good advice of many, I dedicated myself to teleology, an area of philosophy that had been so thoroughly discussed, debated and dismissed it seemed nothing more could possibly be said. But they were wrong, and I eventually found others like me who were beginning to reinvent one of the oldest ways of understanding the world and our roles in it.

Human purpose is a specific type of a more general purpose in nature. Both can be defined abstractly and generally as *forms of self-organized adaptation*. But this definition is not intended as an explanation. The structuring process that we call "self-organization" still needs to be understood. With this book, I hope that I can offer some general insights, showing how chance and constraints might work together to make us purposeful beings. What we learn about our own purposeful behavior will help us understand how nature, society, or culture can be said to act purposefully too.

When you set something up as an object to be gained, as Merriam-Webster defines human "purpose," you do so to maintain and/or improve your "self," whatever that is to you. And your self is, reciprocally, the epitome of all your past desires and accomplishments. If *you* are capable of acting purposefully that means that you (your purposes) are in some sense the *cause* of the action.

Acting purposefully cannot mean following someone else's ideas. Only your own purposes—which arise from your own self-organizing biological and semiotic processes—make *you* a purposeful being. Nothing is purposeful that is the puppet of some other force. To be purposeful is not to be a tool. Years ago, just starting to wonder seriously about these issues, I confronted the usual kinds of questions, What is life for? What is the purpose of life? Or, posed a bit differently, What is the meaning of life? as if it were a kind of allegory, and purposeful beings always served the purposes of someone or something else. I can't remember when exactly I stopped thinking about purpose in this way, but I have, completely. Over the dozen or so years that I've been studying the history of this problem, I have come to realize that purposes can be defined only in relation to the self in question. Your purposes, for instance, are always related to what sustains or furthers your values, what coheres with your personality, and, importantly, what helps you survive, evolve or adapt.

The question of your having a "higher purpose" would pertain to the role that you may have as a part of a larger society or ecosystem. We all play those roles too, as organs not tools. And as such, we preserve our own autonomy. "Organ" comes from the Greek *organon,* meaning "tool" or "instrument," a somewhat unfortunate etymology for an organ is different: it helps create and is created by the individual in which it exists. Tools don't do that. As a purposeful member of a society, for instance, you help create and are created by the society.

Theologians throughout history have made innumerable attempts, some valiant, to explain how people can have free will even if there is a God that determines everything in advance, a God who has a higher purpose under which we are bound, a God who has created us as (effectively" non-organic" in the sense of "not co-creating") instruments of his divine Plan. It cannot be done. Theologians throughout history have tried to co-opt teleology for their own religions. It cannot be done. Teleology is not theology. Teleology comes closer to a transcendental way of animating nature and recognizing some kind of proto-intelligence and creativity *in events themselves* rather than attributing their organization to a Being in control of nature. I say, *comes closer to* because it does not go that far or quite in that direction. Teleology seeks naturalistic explanation for real, natural phenomena. Nature is, as we are, *self*-organizing.

As the new millennium began, I, bravely or naïvely, committed myself to this discredited branch of philosophy, officially submitting "teleological narratives" as my dissertation topic. Although I was working on a doctorate in *English* at City University New York (CUNY) Graduate School, I needed a scientific advisor on my dissertation committee because teleology and biological self-organization are so entwined. By chance one of my literature advisors, Angus Fletcher, had just retired to Santa Fe, New Mexico with his wife Michelle. I would travel great distances to meet with them under any circumstances, but what made me even more willing to visit Santa Fe was the fact that the Santa Fe Institute, perched upon the high desert hills just outside of that city, is the premier center for the complexity sciences, a field in which teleology might be seen anew. The Fletcher's move helped me decide to propose my ideas to theoretical physicist James P. Crutchfield, one of the original investigators of deterministic chaos, who, I had learned from James Gleick's popular book *Chaos*, was interested in the intersections between science and art.

It was a late summer day as I got into my taxicab to go up the hill to the Institute to meet with Professor Crutchfield. As I did so, I noticed something light brown on the red-earth driveway of my adobe hotel: a very large moth with enormous "eye spots," on its wings, which Darwinists claim evolved because they mimic owl's eyes and frighten off predators. Deciding to take it along with me on my interview, I gently nudged it, and it crawled onto my hand and calmly perched upon my palm. It was a providential find, as I wanted to talk to Crutchfield about my dissertation research on a non-Darwinian theory of evolution posed by Russian-American novelist/lepidopterist Vladimir Nabokov in the 1940s. I had become interested in the way Nabokov's theory of the evolution of butterflies and moths corresponded with his theory of creativity generally and of the poetic structure of his own novels specifically.

When I arrived at the Institute, I was greeted by Crutchfield. The image of a late 20th century scientist, fit and suntanned, he sported a ponytail and wore shorts, looking very much like the former surfer he is. We sat on a terrace overlooking the juniper covered hills and began to talk about art and the complexity sciences. Finally, he asked about the moth, which for some time had been sitting on my lap, very slowly beating its huge velvet wings. I told him about Nabokov's work. He was stunned. That summer he was leading a conference on non-Darwinian evolutionary theory, which was thought to be a radically new way of thinking. And here was a student of literature with similar interests. I had had no idea that Crutchfield was working on evolution, and I was just as surprised as he was. We started collaborating right away, and within months I had published my first paper advancing a testable scientific theory, albeit a rather modest one, on the teleological mechanisms behind insect mimicry.

Crutchfield came to be my mentor for several years, and I profited much from observing his tough, empirically-minded approach to the complexity sciences. Although at first his work was simply beyond my ability to understand, I noted that at conferences and workshops when a difficult question was thrown on the table, the entire group would spontaneously turn to him, sitting silent and watchful. His answers were usually greeted with approving noises and pleasant surprise. So I trusted him; when you're just starting to learn you have to put your trust somewhere, and the signs were good. I was not disappointed. He understands the sort of things scientists are not supposed to understand, namely creativity and chance.

I relate the details of our first meeting because the story itself flirts with teleology. My finding the moth by chance helped me to achieve my purpose, and it launched me on a wonderful and exciting journey in science. It was as if Nature had intended it.

Telos is Greek for an "end" or function, which helps explain *why* something exists or why its previous actions occurred: *in order to serve that function.*[1] Telic action requires a *representation* of the goal that helps achieve it. In short, teleologists argue that ideas, or something *like* mental concepts or thoughts, cause events in a way wholly different from the way that objects cause events—atoms, molecules or larger bodies hitting each other and/or reacting (see, for example, Adams, Ducasse, Ehring, Papineau, Taylor, Wimsatt, Woodfield, and Wright).

Assuming that the universe cannot think (and there is no god to do the thinking for it), philosophers and scientists first ruled out *telos* as a cause in the processes of inanimate nature, which, it was believed, simply proceeded mechanically, according to the laws of physics. For a while it was still accepted that *people* (if not all animals) did act purposefully, that is, did and said things in part caused by the representations of goals, in a non-mechanistic way somehow beyond the known laws of physics. But as time went on, it was argued more and more that even mental acts and thoughts were strictly mechanical events—electro-chemical interactions in gray matter. Accordingly, purposeful animal behavior was also eventually pretty much ruled out by science, and the final nail was hammered into the coffin lid of teleology. How could one argue that nature was purposeful if even humans were not?

This situation led to all sorts of arguments about the nature of human responsibility. How could one be blamed for or take credit for any thing one did, if every action was determined by the laws of physics and was effectively

1. In this work I use *telos* as a non-countable noun, contrary to convention and, some will say, correct grammar. Normally one would say, the *telos* or a *telos*. Non-countable nouns are nouns that do not take a definite article such as "a" or "the" and do not have plural forms. Some examples of words used as non-countable nouns are: aliveness, justice, gravity, luck, intelligence, progress, art, and science. In my mind *telos* is not a definite, countable thing: it is an abstraction, a general concept. Though most philosophers agree that "ends" are always general, never particular, their grammar does not reflect this belief and therefore can, I think, encourage confusion. I want my grammar to reflect my conception, so I'm reinterpreting the rules to suit my purposes.

inescapable? I will not be taking up such an unwieldy issue in this book.[2] My aims are more modest. Instead I will explore a related issue: not ethical responsibility but aesthetic responsibility. How are new meanings and forms created? And who or what process is responsible for the new creation?

Art and teleology are intimately related. This is so because teleology involves representation, design, and meaning. Perhaps aesthetics and teleology are actually the same formal discipline. To say nature is teleological is to say nature works like an artist. To say something is a work of art is to say it is teleological. This helps explain why representation, design and meaning (in other words, art) came under attack with the "demise" of teleology.

In the late 20th Century, artists and writers, who had always been creators of worlds, were no longer considered responsible for their actions, creations, or meanings. However original they might seem, what they did was just the inevitable product of their interacting matter. With the supposed end of artistic intention, the romance that used to attend the artist walked out of the cold-water flat one morning without leaving a note.

Art was replaced with something that was called "art" but was not artful in the teleological sense. Paradoxically and with great confusion, the (then apparent) end of teleology redefined what kind of "art" could be considered "intellectual," "sophisticated," or "artistic" by critics, students, academics, and professionals. Purpose was associated with religious narratives and with superstition. It became unfashionable to represent the world in a teleological way. In a teleological narrative, all the events depicted, or at least the key ones, are chosen and included because of the way they reflect, refract, or prefigure a general theme of the story or the end of the story, the resolution of a problem. There is usually progression or development. Events exist in the story *because of* the purpose they serve. 20th Century non-teleological so-called artists decided instead that "realistic" representation should capture a world in which the parts did not relate to a whole. Characters in non-teleological novels wander aimlessly and seldom undergo change, either for better or worse. Many things happen to them that don't seem to add up to anything. Some so-called artists thought that indeterminacy at the quantum level disproved teleology so they set about making worlds that mimicked the true "reality" of the quantum world with its inherent unpredictability and chaos.

---

2. See Alicia Juarerro's *Dynamics in Action: Intentional Behavior as a Complex System* for an excellent treatment of this question.

In 20th Century "arts," accordingly, if there was to be any organization in a work at all, it was appropriate to leave up to the reader, viewer, or critic to "create" it by imposing his/her own interpretation, tinged, of course, with knowing and delicate irony. The reader became the *author*ity. In the visual "arts," the idea that the artist should try to control interpretation or try to plan or design a composition was considered gauche. Painters started representing things *thrown* together, not *put* together for reasons. Intentional organization was out. The word "organization," which has "organ" in it, is virtually synonymous with functional design, teleological design. Many, many visual "artists" abandoned perspective, color theory, and composition. Some writers, particularly poets, not only abandoned grammar and syntax but also semantics. Representation itself—its aims, its uses, its good—was called into question.

Teleological representation was wrongly associated with dogmatic values and morals, with religion, and, during the cold war era, with propaganda. Even the pure formalism of abstract Modern art, which was supposed to represent and abstract something about the world, eventually began to be discounted. This led to the popularity of non-representational art, which, it so happens, is more difficult to judge than representational art because, well, it doesn't represent anything; there is no basis for assigning value. Savvy capitalists seized upon the opportunity and began aggressively promoting value-free art, art that could not be judged according to the amount of representational skill necessary to create it. Something called the "art world," made up of the products of MFA mills, glossy magazines, and pretentious galleries, emerged with its own arbitrary and completely conventional—that is, ungrounded—language. Anything the art world said was "good" was good, even if it was interchangeable with art outside the art world, which was said to be "bad." In this way gallery owners were able to print their own money, as it were. Fake sales of an artist's work drove up the price tag of all his or her work. Of course, the gallery had already contracted to purchase all the artist's work at a small fraction of what they would sell it for. During the 1980s when these sort of abuses reached a height, skilled artists who realized what was going on created a slogan targeted at those who had been lured by the promise of becoming an art celebrity: "It's the dealer, stupid," acknowledged that it was the dealer who was the real artist, a money artist.

Doubtless there are all sorts of other reasons—political, economic, social, philosophical, and aesthetic—for the way the art and literary worlds developed in the 20th century. Most people active in the arts were probably pawns of the economic machinery (even as they helped create and maintain it) and blissfully unaware of the debate about teleology and intentionality. I suspect that most just learned to copy a popular style. Anxious-to-please students of art and literature probably only repeated what they had heard, slogans such as "process is preferable to product," without really knowing anything about teleology or that it had been associated with "product" and very wrongly disassociated from "process." These values and prejudices were known generally as "postmodernism" and "deconstruction," conveniently ill-defined labels that denoted what the important people were saying was sophisticated, smart, and stylish.[3]

I will speak plainly. I do not like non-teleological art. Not only do I think it inexcusably boring, I think it false. Nature *is* a work of art, and there's no good reason why art, representing nature, should not appear artistic, by which I mean intentional and purposeful.

I am not an advocate of Intelligent Design, which, insofar as I understand it from what very little I have read, seems just a roundabout attempt to argue for a supernatural Creator. There are no gods in my cosmos. Nature *is* creative, but without a Creator. Nature is *self*-organizing. When I say "creative," I mean progressively more able to make more complex and astounding things, like us, not quite by pure accident, but by availing itself, in the way that artists do, of the emergent ordering tendencies of chance.

Although *telos* has been variously interpreted throughout history, I make the argument throughout this book that it has consistently involved chance. This goes directly against the predominant grain of contemporary thought that associates teleology with rigidity. While I admit I refine *telos,* as I will show, my definition is implied, latent or prefigured, in every definition from Aristotle's onward.

3. The chief sources from which postmodern thought draws its conclusions about science, generally understood, are: firstly, Thomas Kuhn, *The Structure of Scientific Revolutions*, whose argument that the ruling paradigm determines what science considers true has been misused by many postmodernists to defend the idea that science is "just another narrative"; secondly and significantly Jean-François Lyotard, *The Postmodern Condition: A Report on Knowledge.* Lyotard argues, for example, that because "Quantum theory and microphysics require a ... radical revision of the idea of a continuous and predictable path. The quest for precision is ... limited ... by the very nature of matter," and therefore knowledge itself "resolves into a multiplicity of absolutely incompatible statements" (56-57).

To say nature is a work of art (*sans* Super Artist) requires a new understanding of what it means to act intentionally and of what it means to say that nature's processes are teleological. Fortunately for me, there is at hand a burgeoning field of research that sees the issue completely anew. This "field" is actually a collection of loosely interacting individuals that work under various names: complexity scientists, neutral evolutionary theorists, emergentists, complexity neuroscientists, systems theorists, synergetic researchers, cyberneticists, and biosemioticians. Don't be put off by these labels. I intend to keep this as non-technical as possible.

I want to speak to generally educated audiences—people outside the circle of academic obscurantists—to those of you who get annoyed at disjointed movies or who are baffled by reports of child-like paintings going at auction for more than you could ever hope to make in a lifetime of hard and thoughtful labor. You no longer have to feel embarrassed that you just don't "appreciate" such things. Obscurity is a tyrant that bullies its critics into silence. The only thing one can say is, *I don't understand*, more or less shouldering the blame. But it's not that you don't get it. Either there's nothing there to get or you did get it, you just didn't think it was interesting enough to be the "it" everyone is so excited about. So you said nothing. We've all been respectful; let others have their opinions and tastes. I've been guilty of this too, working uncomplainingly in the arts now as I have for fifteen years. But I'm breaking that silence now. Art, I think, is our most precious resource, our source of inventiveness and our means of progress, and we cannot afford to let it sink into dullness (as Alexander Pope imagined) out of a self-damning respect for other people's opinions and a polite reluctance to say to an artist, But your work is indistinguishable from work that requires neither skill, talent, serious thought, nor effort (other than tedious)!

With this book I will likely offend some of my own colleagues and (very good) friends who enjoy postmodern irony and like very, very "difficult" work that you can't really understand. They will defend the aesthetic experience of "bewilderment" and "confusion" which, I happen to agree, is an essential precondition of poetic experience. But I think it's the *making sense* of the world anew which is truly poetic, not the destruction of sense. *Poesis*, from which "poetry" is derived, means *making* not unmaking.

That's about as much as I intend to say about the movements known loosely as "postmodern." To recount and rebuff their ideas here in detail would do my reader a great disservice since they are obscure and boring—infamously so—and I want a larger audience than they invite. My topic is difficult enough without adding more difficulty to it. Another reason I will not be offering a more than a cursory summary of the criticism is because, to be quite honest, I have not spent the time everyone claims is required to get it completely. (How convenient for its practitioners for these works are generally long and far too dull to finish.) Oh but I know the theory well enough to know it's not worth the time. I passed the required graduate exams. I read Jacques Derrida's *Of Grammatology* to page 200 or so, at which point I hurled it across the room. It was clear enough he had got teleology and Charles Sanders Peirce, an American Pragmatist philosopher and semiotician, on whom he had based some of his ideas, wrong (see Short, *Perice's Theory* 45-59). I am glad that I couldn't get through that book, glad I didn't understand him, and most of all, glad I did not acquire his vocabulary and style. And at the risk of inviting even more criticism from my peers, most of whom have the deepest respect for ideas I cannot even fathom, I admit I know next to nothing about Lacan.

Fortunately, postmodernism is quickly going out of style as I write. There are more than a few who have attempted to salvage it by calling *what I do* postmodern, saying yes, that's what we meant. But I disown you all. The one thing I have in common with postmodernists is, perhaps, a rejection of an essentialist conception of selfhood (*i.e.* a static identity, predetermined by nature). But who doesn't these days? Similarities pretty much end there because they did not attempt a new definition. I, and the others I work with, do. And this is a profound distinction that affects the way we stand in regard to ideas of value and the creation of value.

Postmodernism is being blamed now for much of America's woes, its lack of standards, its nation of drifters. I will not quite go that far, little as I like my Derrida. But I will say that academics in the humanities took such an ironic stance to their own subject that they deconstructed themselves out of respectability. The project has backfired. Few people like us these days. I know. I feel the chill settle at PTA meetings when it is discovered that I am a doctor of philosophy, not medicine. What *use* are we in the humanities? We need to prove to everyone that art and literature are important, as important as science, if not much more so, to understanding the world.

Maybe there's hope that we academics can redeem ourselves yet. While those who have invested their careers in postmodernism are still defending it, the "hip" audience is gone. Organization, form and beauty are making a comeback in the visual arts (due largely to economic causes, not ideas) and no one seems to remember why it was so uncool before. Such is the nature of fads. There is an intellectual vacuum left by the retreating theory and I intend to fill it.

Telos is otherwise known as *final cause,* one of four causes identified by Aristotle's natural philosophy: *Material cause* describes how the physical properties of matter determine what a thing is and how it will react with other things. For example, an ivory ball will roll differently than a wooden ball, as the density and weight of the material determines how much resistance it has. *Efficient cause* describes how the agent (person, animal, or even a moving object like a billiard ball) acting on something determines what happens. For example, the pool player, the cue stick or ball hitting another ball at rest is the efficient cause of the latter's moving. *Formal cause* describes how the "blueprint" or the natural laws of form determine what can be. Some forms are physically impossible; others are very probable. Experienced pool players have learned that certain types of moves can be expected to result in certain types of outcomes, and they may apply their knowledge of geometry to their game. *Final cause* describes how the "end," or the function something ultimately serves, determines what happens or why something develops. The ball was struck *so that* the pool player might win the game and further develop his abilities and reputation.

These causes can be applied to processes and things in nature whose efficient cause is not a person, as it is in a pool game. For example two of the material causes of a plant are cellulose and chlorophyll. Some of the efficient causes acting on the plant, causing it to move or change, may be sunlight, water or wind. The formal cause is the type of flower, bush or tree it will grow into, and the structure of a plant does indeed follow geometric principles for proportion and efficiency. The final cause of the plant may be to be beautiful to bees, bear fruit for seed-distributing birds, give off oxygen, process carbon dioxide, or whatever it might do in order to preserve itself, make more of itself or adapt to work better.

Only the first two kinds of causes, material and efficient, are used by science these days to describe natural processes. Although some might argue that natural selection *is* final cause (Ruth Millikan, for instance, naturalizes teleology/intentionality by explaining it in terms of natural selection) most

Neo-Darwinists believe *On the Origin of the Species* replaced teleology with a theory that deals only with physical (material or efficient) causes. Contemporary science has a difficult time understanding how types of formal designs or types of purposes can affect things. This difficulty is quite understandable as types are idea-like.

In teleology, I should note, formal cause and final cause are often confused and conflated. There are good reasons for this, which I will address in a later chapter. Suffice it here to say that formal cause is behind the creation of organization, making the parts of a system work together in a harmonious whole. Only when the whole goes beyond merely creating and maintaining itself and it *adapts* in response to something outside itself, is it fully teleological and spurred by final cause as well. It must spontaneously form and also *change* fortuitously. It's also important to remember, as Aristotle himself pointed out, that all telic phenomena involve not only formal and final causation, but material and efficient causation as well.

Final cause is sometimes called *reverse* cause because it seems the ultimate purpose the end state happens to serve is supposed, somehow, to determine its beginning. For instance, a pseudo-teleologist might say, "The first brightly colored flower mutation occurred in order to attract bees." That is not the way I will be defining final cause here. The future does not affect the present. The "future" is just a word we have for something that doesn't actually ever exist, and *telos* certainly cannot be "in" the future.

But a "whole" does exist, and it does affect the parts, and wholes, as such, do have effects in the larger world. A flower is made of parts that interact forming the whole, and the whole reciprocally constrains the parts. A flower's matter and form help preserve and further the flower. When I talk about organization and design, I am talking about the functional arrangement of parts to a whole. This is what I mean by "end." I do not mean the end of a sequence in time that is supposed to affect the beginning. The "end" for a teleologist is a whole that affects the parts. *The whole is the type of result brought about by the interacting parts, and it is the type of result that allows the parts to continue.* There is no time travel involved in a teleologist's life, and we never try to get around the laws of physics which, we pretty much agree, work in one temporal direction.

NOT "END-STATE" BUT "EMERGENT WHOLE"

A "whole" is defined as something that is other than the sum of its parts. For instance, a human being is made up of molecules that each in themselves behave predictably enough, but the human as a whole sometimes behaves in ways that no scientist or even supreme intelligence could predict. We say human actions and/or personalities are *emergent.* To a lesser extent, the development and/or adaptations of plants are emergent. We cannot predict exactly how a flower will develop, and we cannot predict how its species might adapt and evolve.

A "whole" is kind of dynamic entity that by constantly changing its parts remains more or less the same. Humanness. Flowerness. There is no one static definition of any whole; it is a class or category with fluid boundaries. If a whole is different from the sum of its parts, if it is something fluid and indefinable that doesn't have quite the same kind of "thingness" its material parts seem to have, then a "whole," as such, is not something that any other thing can directly interact with. We can only interact with or "know" traces of a whole, you might say. Signs of it. In fact, we can only theorize that wholes exist at all because of their otherwise inexplicable effects we observe in their parts (or a sample of their parts' actions). For example,

- Molecules that are part of a systemic whole behavior differently, in a more constrained way, than those same molecules would if they were free.
- You can tell something of the wholeness of a person by some of his or her actions, which are usually more or less characteristic or that person. People have a wealth of options available to them, but they only consider a small sample, a sample that reflects the character and past of the person.
- Individuals that are organized together and interacting, like birds in a flock, tend to behave in ways that are more orderly than they would if they were separate. The parts of a whole are constrained by that whole.
- A species is a concept that is impossible to define exactly and there is no pre-existing essential nature of, say, a human being. But we can infer that species exist because offspring conform to a general type that emerges in the process of development and evolution.

We can say, then, that the whole is *represented* (or known) in the relatively limited or orderly behavior of its parts.

It is in this way that I bring in the notion, mentioned above, that something like mental concepts are involved in teleological phenomena. I came to think of the constrained behavior of the parts as a *sign* of the whole through Crutchfield's influence. He claims the limited or more regular behavior of a part of a complex system is a "model" of the whole. The model or sign is idea-like because signs are not material things but relationships between things. *If a sign relation is involved in the causal process, then we have a different kind of causality than efficient or material causality.*[4]

My view can be considered a *biosemiotic* view of purpose. Of all the different research fields currently investigating intentionality and teleology, I think biosemiotics will be the most fruitful. It's a fairly new discipline and may sound rather odd since semiosis is usually associated with human interpretation, not with biology *per se*. But biosemioticians do not commit the sin of anthropomorphism: they argue that conscious human semiotics should be studied in terms of the more basic forms of signaling, such as found in cells and simple organisms—not the other way round. Although human language is an astounding achievement, it cannot be wholly unlike what gave rise to it. There must be some more primitive semiotic processes from which it evolved. To argue otherwise would be to assume an indefensible humanism that makes us a little too miraculously unlike our environments.

Non-human or natural signs can be found in any *nested self-organized system*. To explain what I mean by this, I will look at James Lovelock's notion that our ecosystem is a self-organizing entity called "Gaia" that purposely regulates itself. Once considered merely a fanciful metaphor, Gaia is now accepted by many in the scientific community as valid description of ecological robustness and interaction. I will use Gaia as a test case for the theory of semiotic purpose. This example will also give you a better idea of what I mean by an "emergent whole" being a "teleological end state" or "goal state."

---

4. I believe that every response to something as a sign, *as it happens*, is *poesis,* making meaning, not semiosis per se, which depends upon the idea that a background of meaning already exists. Semiosis proper doesn't emerge until a trend starts forming out of poetic processes. However, responses to things as signs are not purely accidental or *merely* arbitrary: I do not advocate constructionism. Even poetic responses are constrained and somewhat probable. This argument is a long technical one, and I make it in "The Poetics of Purpose." I recommend this paper to postmodernists who won't mind the difficulty and who will probably find the semiosis/not semiosis idea to their liking.

In what Lovelock and Andrew Watson called the "Daisyworld simulation," how biological organisms work together was modeled, much like the parts of a thermostat, to maintain a particular overall temperature range (*i.e.* a *type* of temperature). Our ecosystem does this on its own, without an engineer, and so it is self-organizing, not mechanistic as a thermostat is. Lovelock stripped down the biological thermostat model to two parts: black daisies, a type of plant that can grow in cool temperatures whose color happens to absorb heat, thus slightly warming its surroundings, and white daisies, which are able to grow once the black daisies have become sufficient in number to warm the environment, and which reflect heat away, thus cooling their surroundings.

When it gets sufficiently hot for white daisies to appear, the black daisies continue to flourish and increase the temperature. Finally, it gets too hot for black daisies, and while the relatively few white daisies do fine, the black daisies begin to diminish. As the black daisies diminish, their doing so is a sign, we may say, of the holistic temperature, which is too high for them. Their diminishing *indicates* high temperature (an "index" is a type of sign in semiotics terminology). As the now rapidly growing heat-loving white daisies cool down the planet, the black daisy growth rate may pick back up, while the white daisies' rate begins to slow. The white daisies' and the black daisies' growth rates are constrained by the whole temperature, which they help create.

A temperature equilibrium is reached which sustains both types of daisies, and we may say that black and white daisies have synchronized to become part of a holistic emergent system—stable temperature. We can argue that "Gaia" exists as a stable regulating entity because we can see signs of this "system" in the constrained daisy growth rates.

Of course, our real ecosystem is infinitely more complex than Daisyworld, balancing numerous differences: chemical gradients, pressure gradients, and *etc.* all interacting to form extremely robust balances. Many people believe that Gaia is a purposeful being. That's a little too far-fetched for me. I say Gaia is a self-organizing entity that exhibits proto-purpose, and can be categorized with other inanimate self-organizing systems, like tornadoes, autocatalytic chemical reactions, and so forth. Were Gaia, as a whole, able to harness the stable tendencies of an *other* self-organized entity (another planet, say) to her advantage resulting in an adaptation, then I would be more open to thinking of Gaia as a purposeful being.[5] A fully purposeful entity, in my restrictive view, doesn't just maintain itself; it has to *do* something of value to itself in the world. The biological individual is the best example of fully purposeful entity, but I am open to other arguments, say, for example, the argument that groups can be considered purposeful entities.

The inclusion of semiosis (the study of signs) and incorporating semiosis into complexity science's theory of self-organizing wholes distinguishes my argument for purpose (or free will) from, for example, that of Compatibilists, like Daniel Dennett, *and* from that of their opposition, Libertarians. This is not the usual argument for free will. This is a relatively unknown and new argument, but, as I say, not entirely original. Gregory Bateson, Jesper Hoffmeyer, Alicia Juarrero, Stanley Salthe, Robert Ulanowicz, and others have been doing similar work in this area for thirty years or more, and it has even deeper roots in ancient holistic thinking.

I don't want all this talk about holism to align me with new age spiritualism, crystology, and ambient electronic music. My kind of holism comes out of the complexity sciences. No doubt there are incense burners among complexity scientists, but perhaps no more than in any other academic discipline. In this book, I promise I will try to demystify *telos* without killing its beauty or voiding everything about it that makes it a special kind of cause. The health of teleology today depends upon finding a scientific understanding of "wholeness," not a warm fuzzy one, and a clearer understanding of "representation," of what it means to be a sign of something that is beyond complete description.

5. Likewise, Francisco Varela's autopoietic systems are not fully teleological, in my view, as they only preserve their own structures, differentiating themselves from the external world, and there is no real way to describe a process of adaptive change and creativity using Varela's model. This issue will be explored in a later chapter on "Directionality and Originality." I want to mention in this context that Robert Ulanowicz compares a biological system to chemical autocatalysis (in both, parts create and sustain the whole). He notes that only the parts of biological systems are able to *adapt* to make the self-sustaining process even more efficient. So insofar as Gaia adapts, it might be more purposeful than inanimate chemical autocatalysis could be, yet still not fully purposeful in my view. Various terms have been introduced over the years to describe gradations of purpose, *e.g.* "teleomony" is said to involve the apparent purposefulness of evolved organisms whose self-sustaining activities are believed by some to be determined genetically. Norbert Wiener's mechanistic view of purpose is telemonic, because either inherited (in organisms) or designed by an engineer (in smart technology). Real (human) purpose, in contrast, is said to be caused by "choice." However, when one considers that "choice" is not sufficiently explained or described, we must accept that our purposeful behavior is much more similar to this "apparent" purpose that is only "telemonic." And so all "kinds" of purpose require a more general description. This is what I'm trying to get at with this book.

Today teleology may not be popular with scientists or postmodernists, but the idea of *telos,* if not the term itself, is still widely used elsewhere. A number of distinct species of the term have evolved since Aristotle attempted to describe final causality. Christians associate *telos* with luck, good and bad, known to them as the mysterious ways of Providence. To engineers the *telos* of a thing or event might refer to the mechanistic unfolding of a predetermined design, and would be limited to things artificial. Panpsychics and vitalists (insofar as they're still around) think *telos* is the immaterial cause of aliveness. In the last few decades, *telos* has even come to be associated with the second law and the production of entropy.

Contrary to popular belief, there are many different kinds of teleologies: there is Aristotle's version, Christian versions, Kantian versions and Pragmatic versions, to name those that I will examine in later chapters. Teleologists arguing with non-teleologists often aren't working with the same definition of teleology, and consequently, they are not strictly "arguing" so much as talking past each other. I hope to clarify things a bit by making distinctions between different types of teleologies.

The fact that *telos* can refer to several apparently contradictory ideas signals a hidden richness. Perennially dismissed by empiricists, evolutionists, and postmodernists, *telos* keeps returning like a misunderstood ghost. Teleology demands a clearer conceptualization, and it will have it in one that reconciles all this apparently contradictory ideas into a musical coherence. There is a new way of thinking of nature as acting purposefully (contra Charles Darwin) and of ourselves as being capable of "free will" or rather being "self-determined," to be more precise (contra some implications of Newton's reductionism). Purpose, in this new view, is emergent.

I suggest that "emergent" can replace "postmodern" as the name for our next new cultural era, which will, I hope, bring us art that's meaningful and adds something to what we know about ourselves and our world. Robert Laughlin's book *A Different Universe* (2005) makes the case for a paradigm shift in physics to emergentism. Robert Reid's *Biological Emergences* (2007) makes the case for a similar shift in evolutionary theory, demoting NeoDarwinists to the task of explaining stability in species, and pronouncing them incapable of explaining creative change or progress. In this work, I make the case for an equally profound shift in arts and literature.

Chapter 2

CHANCE AND PURPOSE

Poetry, even that of the loftiest, and, seemingly, that of the wildest odes, [has] a logic of its own, as severe as that of science; and more difficult, because more subtle, more complex, and dependent on more, and more fugitive causes.

– *Coleridge*

Why did I veer off on a career path so untrodden of late? And given my interest in science and empiricism, what inspired me to look into an area of study associated with religion and superstition? My story is a bit like that of Saul of Tarsus. Saul, who became Paul the cornerstone of the church, had spent his early days persecuting Christians and had a conversion on the road to Damascus when he encountered a vision of God. My conversion wasn't quite so dramatic or in any way miraculous, but I was, at one time, very much against teleology. As an undergraduate, I became an atheist and joined a secular activist group. I was also an avid Darwinist, and I would mark-out the teleological language in my biology textbooks. For instance, I might change, "Birds have feathers *in order to* keep them warm," to "Birds have feathers *which happen to* keep them warm." With such corrections, chance always replaced *telos*.

Then one day it occurred to me that "chance" was just a term used to denote "not caused by material or efficient causes," which logically might well mean, "caused by some other kind of cause," perhaps final cause. In fact, functional coincidences are the only "evidence" of final cause (or divine intervention) that has ever been offered by its defenders. Applying Occam's razor, we may say that chance (whatever that is, for we will need to define it) is final cause. I will define it below; suffice it to say now that chance, as I use it, which is to say *meaningful* chance, is a particular kind of selection process involving constraints and feedback.

My sudden conversion occurred when I, at the time a wannabe postmodernist, was trying, between college semesters, to write a novel without any trace of an Author behind its details and descriptions, trying hard not to add artful patterns to the events but to make everything very literal and "realistic." I wasn't very successful at this. Although I had chosen my characters' names randomly, their names came to reflect the type of role they played. The novel was partly set in Japan, so quite predictably there was one male character called Hiro. I hadn't planned a significant role for him, but he became the hero, albeit an ironic one. At the time I was studying German, and one of the male names I chose, initially just as a placeholder, was "Gottlieb," a common name used in my conjugation exercises. "Gottlieb" happens to translate as "God love." Usually, only foreign speakers of a language tend to notice the literal meaning of names. For instance, I don't much think of myself as victorious in any way. But there developed around this character a critique of the practice, common to some forms of poetry, of intermingling of religious devotion and romantic love. I had another character who happened to be in the shipping industry, a common trade for Japanese, and the entire story grew into a large metaphor for the voyage out, as it were.[1] I discovered that themes developed despite my intentions. Then I realized, as all artists eventually do, that these were my intentions. This is what intentionality is: the *emergence* of meaningful patterned behavior, and the emergence of an author as a coherent self. As Czech novelist Milan Kundera says, "You cannot escape your life's theme."

What I mean by this is that, as a novelist, there are certain kinds of patterns and ideas that I tend to notice more than others, certain themes that appeal to me because they are useful to the extent that I can fit them into other themes. Once a theme occurs once, its chance of occurring again is increased, that is, once the theme of religion has crossed my mind, there is a greater chance that I will be reminded of that theme by some other random detail. I tend to think thematically. I don't have to do so consciously. This is simply the way (most) brains work whether we will or want them to or not.

Now it may seem to some that conscious deliberation is necessary for intentional actions or "free will" to be exercised. Although intentionality and free will are related terms, I prefer the former in this analysis because it does not necessarily imply, as "free will" usually does, conscious decision-making. Being conscious of one's purpose is not a requirement in my view of

1. That novel became *Smoking Hopes.* "Hopes" is a brand of Japanese cigarettes.

purposeful behavior. Contemporary neuroscience corroborates the idea that self-organizing processes in the brain give rise to thoughts/responses, and we become aware of thoughts and actions *only after* they have already been "chosen" consciously, that is, after the electro-chemical "message" has already been sent (for a review see Wegner). According to my notion of purpose, trends and patterns of behavior define selfhood, and those actions and thoughts that are the products of a constraining, defining selfhood constitute purposeful or intentional behavior. So the patterns that emerge from our unconscious actions would be just as much "us" as those that emerge from what we perceive as our "choices." In either case, self-organizing process are responsible.[2] "Chosen" actions just have the added feature that we are aware of them as "our" actions sooner rather than later.

Purposeful behavior then is defined by the self-organizing tendencies of a selection process that does not involve conscious selection. I later realized that "themes" develop in nature too through random processes not too dissimilar from the artistic process described above, when there are constraints and feedback. In nature, as in the artist's mind, when one thing is coincidentally near another, or when one thing is coincidentally like another, this may affect or constrain the outcome or the way they interact. This is selection, but not á la Darwin, not quite. It is not selection for reproductive fitness. This kind of selection simply builds formal patterns. The "by-chance" near become like each other through habitual interaction, and the "by-chance" like bond together, forming constrained systems. (A "system" here is just an organized group whose parts interact to form a coherent behavior or cycle.)

In nature, mutations (genetic or other) are generally concordant with the original configurations, very much as with meaningful creative change in art, which is largely metaphoric (similar) or metonymic (contiguous). The repeated production of similar and/or contiguous changes (rather than merely random changes or changes reflecting a much wider range of possibilities) automatically *results in structural patterns and associations*, the basic building blocks for the creation of systems and sub-systems, that is, for the creation of life, art and language. Look at it this way: if genetic mutations usually

2. Alicia Juarrero notes in "Top-Down Causation and Autonomy in Complex Systems" that the constraints of a whole, *i.e* the self, exist at a different temporal phase than the phase in which the actions of the parts take place. Thus even though decisions are made and we only become aware of them later, our selves are always already pre-existing and constraining decisions—and causing them but not by efficient cause.

result, not in some random group of dissimilar cells all mixed together, but in a group of cells that form some kind of pattern containing different types that might interact and feedback the way black and white daisies do, then mutations might result naturally in a new stable "system" that might produce an effect, as the daisies produced a stable temperature, that might be useful to the organism in some way. *Random things aren't useful in the way that things that change in fairly predictable ways are.* Such new forms can emerge without natural selection, which only helps them proliferate and/or stabilizes them in a population.

This kind of process is selection for self-creation and self-maintenance of systems or entities, which we can think of as more or less *functionally neutral* pattern building and development of stable tendencies, and which must occur prior to natural selection. It is a *formal* selection process that creates the entities, which natural selection can later favor or not with respect to others with which they compete. A self-organized entity might be an autocatalytic chemical reaction or gradient reducing cycle, any system that forms spontaneously when there is some difference, like the temperature difference of Daisyworld. When these types of self-organized entities are harnessed by life in metabolic processes, allowing the organism to better survive and to develop, there is purpose.

It is because these kinds of creative self-organizing processes are going on in nature, prior to natural selection, that Intelligent Design (ID) people find fault with Darwinism. I think ID people are partly right: an additional explanation is necessary, but a Designer is not the answer.

Such patterns in nature are the emergent *ordering* tendencies of chance (contiguities and similarities), which have tended, in Western history, to be attributed to a supernatural being or to *telos*—or to be denied by science completely. The *disordering* tendencies of chance have gotten all the interest, and chance has tended to be understood simply as relating to the production of entropy. Chance is treated as if it were always a destructive force, what rusts your car and messes up a deck of cards if you drop them. And these things are true enough, but we also need to look at the other side of the flipped coin.

In the early days after these revelations, which changed the course of my career, I had no idea how one might make a scientific argument for emergent self-organization or purposeful behavior using chance. I knew absolutely nothing about neuroscience or how the brain might work in this inescapably teleological way. I was not familiar with the term "self-organization" at the time, although I had experienced the phenomenon in the writing process. All I can say is that I *suspected* that there was, in all of nature, a type of causality that involved the unpredictable effects of analogies (similarities) and coincidental intersections (contiguities) that led to constrained behavior of those near/like parts, and I knew that conscious deliberation was quite irrelevant to these processes (I give a more concrete illustration of formal selection below in the section "Chance"). I also suspected that having a brain would be irrelevant too. Constraints and feedback exist even in self-organizing chemical reactions too and something like representation, as I mentioned earlier, would emerge as the reflection of the whole in the constrained actions of the parts. I would not have been able to put all this quite this way at the time. I only had a strong feeling about it. Such a feeling, of course, might be a symptom of insanity. Or it might not. Let's see.

PURPOSE IS NONLINEAR, NOT LINEAR

Many of those who have been most critical of teleology assume that telic design involves a *conscious* Designer that sets the whole machinery of the universe in clockwork motion according to some predetermined plan. This couldn't be further from the true general feeling of teleology as its been practiced historically. There are two excellent summaries of pre-Darwinian evolutionary biology, which I highly recommend as a general introduction to teleological ideas. If you're interested in further reading, *start here rather than in theology or philosophy or narrative theory*. Timothy Lenoir's book *The Strategy of Life* (1989) rehearses the work of early 19th century German theorists (teleomechanists, structuralists, and morphologists), for example, Karl Ernst von Baer and Johannes Müller. E. S. Russell's long neglected book *Form and Function* (1916) is another compendium of insightful teleological thinking in the life sciences. (I will give my own short history of teleology later, in Part II.) Much of the 19th Century German work was not translated into English, the language in which most of the work in Darwinian evolutionary theory has been done, so it missed the boat, so to say. In the late 19th and early 20th centuries, some teleological ideas did make brief appearances in the work of James Baldwin, C. H. Waddington, Richard Goldschmidt, and D'Arcy Thompson, but they did not make their way into mainstream evolutionary theory.

It's a shame these ideas are not better known as they would help correct what I have found to be a popular misconception: the idea that teleologists use linear causal chains, A→B→C→D to describe purposeful processes. That is, "A happened in order for B to happen in order for C to happen in order for D to happen," with D being the goal or purpose. Anyone who uses this kind of description is a reductionist or theologian in a teleologist's garb. The formulation "A occurs for the sake of D" is also often taken to be the perfect description of teleological explanation. It means "A (a type of behavior) is required for D (a goal as a type of effect) to obtain" and "A's being necessary for D is sufficient for A to occur," or to put it a bit differently, "D's requirement of A tends to bring about A." Such formulations, Charles Taylor's or Larry Wright's for example, invite the criticism that final cause is reverse cause and philosophers have to perform logical summersaults to get around this. Although I sweep away decades of careful thought done by philosophers of great reputation, I am not interested in such formulations. I do not find them helpful. In fact, if I may be so bold, I think they are wrong. Teleological behavior, in my view, is always some kind of self-organized behavior that involves the way the whole constrains the parts, and this fact is not explicitly part of these descriptions.

Those whom I consider "true" teleologists (and these would be some biologists as well as philosophers like Aristotle, Kant, Emerson, Bergson) are more inclined to think in terms of cyclical and reflexive series, A↔B↔C↔B↔A where As and Bs are parts and C is the whole. Organisms, for example, are telic in the sense that their individual cells and organs owe their existences to functional whole in which they develop. As Immanuel Kant put it so well in *Critique of Judgement*, in an organism:

> *every part not only exists* by means of *the other parts, but is thought as existing* for the sake *of the others and the whole—that is as an (organic) instrument….also its parts are all organs reciprocally* producing *one another.... Only a product of such a kind can be called a* natural purpose, *and this because it is an* organized *and* self-organizing being. (220)

My slightly different formula for purposeful behavior is as follows: Individual parts (various As and Bs) interact mechanistically and randomly, and—through formal and/or functional selection processes—form an organized whole (C), which constrains the parts (As and Bs). That whole (C) interacts with and adapts to another entity (X).

One may wonder about my inclusion of an additional system (X), and I will explain more carefully a bit later why fully teleological behavior requires both self-creation/maintenance of C and also adaptation of C through interaction with other entities like/unlike C. Note that the various parts (As and Bs) in this formalism may themselves be systems and sub-systems.

Unlike the formulae of many contemporary teleologists, my formula is not linear. Telic behavior is not like a journey from point A to point Z. In their discussion of basic conceptual metaphors, Lakoff and Turner write,

> *It is virtually unthinkable for any speaker of English (as well as many other languages) to dispense with [the metaphor "purposes are destinations"] for conceptualizing purpose.... To do so would be to change utterly the way we think about goals...* (56)

While Lakoff and Turner may be correct in this assessment, unquestioned devotion to a miscast metaphor does no one good. Changing the way we think about purposeful behavior is precisely what I seek to do. *Purpose cannot be located at a spatial or temporal distance from the self.* Purposeful behavior is not like walking down a road toward a destination. The metaphoric language of goal-directed behavior—a predator chasing prey, an arrow shot at a target—does not suit my view of purpose at all, and I try to avoid it as much as possible.

Ernest Nagel is among those who have concentrated on purposeful *homeostatic* systems. Such systems have *stasis* as the "goal," and homeostasis may be visualized as a circular loop, not a linear chain. One can feel how ill fitting the goal metaphor—which physically separates the self/agent from the goal—is in such circumstances. The maintenance of a tendency or type of response involves feedback and is clearly more circular than linear and therefore requires a more suitable metaphor than "goal-direction," which calls up an image of an arrow flying toward a target.

I define purposeful action (as opposed to mechanical reaction), not in terms of a self pursuing a "goal," but in terms of whether or not "self-causation" obtains. As Alicia Juarrero has noted, these issues were first raised by Aristotle. In attempting to explain purposeful (voluntary) action, he found it necessary to posit an object of desire, a goal, as intentionally represented. He was trying to answer the question: How can a self, as a whole, move itself (or parts of itself) if it is not separate from its parts? How does the mover move itself? His solution was to create an external, immaterial cause, *i.e.* a representation or goal that "caused" the movement. And so we have come to visualize a goal as outside the self, drawing the self on. In the context of critiquing the uses of him made by action theorists (Freeland and Furley), Juarrero glosses Aristotle's solution this way (he was using a lion hunting as an example):

> *... the soul cannot efficiently cause itself, so in reaching out for the gazelle as food [i.e. as significant], the lion's psyche is the unmoved mover of the body. The lion's soul thus requires an external object (the gazelle), as intentionally represented, to actualize the soul's desire by serving as final cause ... or object of desire....* (Dynamics 17-19)

By using a complexity science/biosemiotic approach, I can avoid Aristotle's dilemma, arguing that the self, as "whole," affects its parts by providing constraints. With repeated exposure to each other, the parts are constrained by signs of the whole in each other's actions.

The purposeful acts of chasing, seeking, fleeing and *etc.* are self-organized responses to signs of self. Think of a gazelle as what the lion's body recognizes as part of its metabolic cycle. The gazelle is not the real goal. The gazelle is means of survival, which is the ultimate end. So instead of imagining the teleological process going in a linear direction—the lion (agent) chasing the gazelle (goal)—imagine instead the cycle that maintains the lion's life and the gazelle caught up in that cycle. It is as if the lion's selfhood extends itself into its environment and identifies a part of itself (potential food) and takes it in. The characteristics of the prey becomes a "sign" of the ultimate end of survival when they interact with the predator's evolved repertoire of self-organized responses to the world. The acting of the lion into the environment is not different from what, say, one sub-system of the body might do interacting with some new substance that has just been introduced into the body. The limbs that instinctually begin pursuing the gazelle are not so different from a cell that begins to respond to a nutrient. Chasing is cyclical behavior insofar as it is part of a metabolic cycle.

Only under these conditions of recognition and maintenance of a cycle, can the lion's response, chasing, be considered purposeful.

STASIS AND CHANGE

On the one hand, maintaining a systemic organization or a tendency is archetypical of purposeful behavior: the prime example is eating to survive. On the other hand, maintaining a tendency can also seem robotic if that behavior isn't flexible or can't be altered if it becomes unhealthy. Truly purposeful behavior must be flexible.

Let's consider a person with reactive hypoglycemia, an addiction to glucose, a perversion of our healthy need for sugar, which was hard-to-find in the uncultivated environments of our distant ancestors. The hypoglycemic's system keeps waffling between having too little sugar and having too much. Hungry and tired, the hypoglycemic seeks out and eats too much simple sugar. Then he feels too hyperactive. Insulin rids his blood of excess glucose, depriving his brain of the sugar it needs for healthy function; he feels tired and hungry again and seeks another sugar fix. In order to escape from this unhealthy cycle, the person needs to trick his system into recognizing something other than simple carbohydrates as part of itself—what it needs to maintain itself—and thereby forcing a systemic change and re-self-organization. When the hypoglycemic has low-blood sugar, he really does need sugar fast, but if he can learn to associate the bad sugar—simple carbs—with illness and associate complex carbohydrates with feeling better, he can redefine what the body recognizes as parts of its metabolic cycle for survival. Once the hypoglycemic forces his body out of the unhealthy cycle, he will tend to crave healthier food more often.

Purposeful behavior doesn't maintain a single-minded tendency. An organism must continually grow and adapt to keep itself alive and healthy. A person who can change himself for the better seems more purposeful than one who is living with addiction. All purposeful and/or teleological behaviors should be defined in terms of maintaining cycles that are working well and evolving different cycles when the situation requires it.

EMERGENT, NOT DIVINE

Although Kant argued that telic effectuality arose from the *interactions* of the individual parts of a system in a complex web, as Tim Lenoir notes, a few of his followers in early 19th century biology began to look for an actual physical source of the "principle of organization" in a germ cell or a seed. In this view, teleological phenomena would be set in motion by a physically well-defined initial condition, a kind of plan or blueprint. But this just begs the question of what or who made the blueprint. It doesn't solve the problem of explaining how final cause actually works through self-organizing processes.

Jacques Derrida (1930-2004) makes a similar move referring to the "center," a principle of organization that, he claims, may be described "as readily as *arché* as *telos*," that is, as origin *or* end. His critique of *telos* stems from a refusal to conceive of a final cause that does not have an a priori existence, like an instruction manual or codebook or divine plan. The concept of telic organization does not depend, as Derrida argues, on the existence of "a linked chain of determinations of the center" (960). Quite the opposite. *Telos* arises from nonlinear processes. Derrida concludes that teleological phenomena are in fact constructed by those perceiving them. I argue instead, after C. S. Peirce, that they are emergent, not "merely" constructed, and they have a reality apart from our perceptions of them.

It may be true that there were a few so-called teleologists who sought a fixed, stable source for the organizing principle, that unfolds in a linear process, but generally, only those like Derrida making arguments *against* teleology tend to characterize it this way. Genuine teleology seeks to escape reductive analyses. It seeks, in additional to material causes, evidence for an emergent cause *immanent* in the process itself. As Henri Bergson writes in 1907, the flawed notion of a predetermined teleology:

> *implies that things and beings merely realize a programme previously arranged . . . As in the mechanistic hypothesis, here again it is supposed that* all is given. *Finalism thus understood is only inverted mechanism.* (39)

To reiterate then, true teleologies posit neither an agent, nor blueprint, nor a physical code-like "seed" that determines telic behavior. This would beg the question of what is the cause of the agent or the seed. True teleology requires no further explanation of final cause; it emerges in nonlinear systems through interaction and feedback.

Alicia Juarrero (*née* Roqué) may have been first to compare Kant's teleology to nonlinear causality and late 20th century complexity sciences ("Self-Organization"). Our current arguments restate pretty closely Kant's definition of self-organization. Although early teleologists were not precisely correct in their understanding of telic phenomena, they generally had a good intuitive grasp of what was impossible to describe fully with the tools available to them at the time. Now we have better means of understanding nonlinear systems and complex dynamical systems as teleological systems.

NONLINEARITY

A "nonlinear" system is a group of interacting parts that share the same general constraints. It is "complex," not merely complicated, meaning what it does is inherently unpredictable. With a linear system, the more information one gathers about it, the more one's ability to predict its behavior will improve, in proportion to the amount of information gained. With a nonlinear system, in contrast, *one's ability to predict behavior does not increase proportionally the more information one gathers about its parts.* There is something effectual in the dynamics of the system, the functional relationships between the parts, which adds to what it is. So summing up a detailed description of the physical laws guiding all its various parts in isolation does not get you the knowledge you need to make *specific* predictions about how the system, as a whole, will behave.[3]

According to complexity scientists, nonlinear systems, emergent systems, such as human beings, are inherently unpredictable. People are not solely determined by their genes, nor by their environments, *nor by the sum of both*. It's not possible to predict exactly how genes will react with every given environment. Biological genes are not prescriptions, but only the raw materials—with specific kinds of formal constraints, given in roughly the right amounts, in the right sequence—that in the right environments self-organize into a whole. Successful development does not depend upon a fully detailed genetic blueprint but on feedback and the high *probability* of conditions being conducive for a general type of development. As Jesper Hoffmeyer argues, "Living cells through their [filtering] membranes, use DNA to construct the organism, not vice versa" (*Biosemiotics* 32).

---

3. Learning to predict the behavior of complex systems involves, not studying the parts and summing up behavior, but interacting with the system such that the observer becomes part of the system and can model the behavior of the whole. See Crutchfield, "Is Anything Ever New? Considering Emergence."

With the hindsight provided by these new theories of development and evolution, the old teleologists suddenly seem pretty smart. We now can clearly see that teleologists throughout history have typically described *telos* in nature as some sort of internal constraints, definitely *not* as the direct effect of the hand of a supernatural agent. When cultural critic and semiotician Roland Barthes in "The Death of the Author" (1968) argued that the notion of authorial intentionality in narrative is as untenable as teleology in the natural world (as per Nietzsche's "God is dead,") he confused final cause with conscious human effort and the following of plans already formed. The attempt to kill the author missed its mark.

The author in teleological phenomena is an entity that is formed in the process of writing according to the ordering tendencies of chance. In most teleologies the random interactions of the parts of a telic whole occupy, if not the center of the argument, a prominent or distinguished place. As I mentioned already, contrary to what is generally assumed, there are several varieties of teleology. *Every time there have been changes in theories of chance, teleology has responded.* In Part III of this book, I will identify different types of teleology: Aristotelian, Analogical Determinism (after St. Augustine), Deterministic Fortuity (after Kant), and Pragmatism (after Peirce), and will show how each teleology actually turns on a new theory of chance. Theorists writing about artistic creation have likewise always noted the role of chance in the form of inspiration and spontaneous self-organization. Teleology and intentionality are not bound up with the classical notion of domino causality or "linear" determinism, but nonlinearity and chance.

CHANCE

It's time to define "chance," a slippery term that plays a key role in this work. In ordinary language the term "chance" is often used in opposition to purpose, but other times it can refer to the precondition for purpose. Sometimes it is almost synonymous with purpose.

Here are some of the very different ways "chance" is used in ordinary language:

1. "I found my lost watch by chance when I happened to notice it lying on the ground." Here "by chance" is opposed to "purposefully."

2. "There is a small chance that it will rain tonight." Here "chance" is a measure of ignorance about the exact probability of occurrence of the conditions being right for rain at a particular time.
3. "The volcano blew by chance just as the helicopter passed over it." Here again as in the first example, "chance" refers to coincidence that was not intended.
4. "My winning the lottery was caused by chance." Here "by chance" could mean a.) some kind of causal force like "good luck" or it could mean b.) the coincidence of the winner having picked certain numbers and a machine having drawn the same.
5. "There is no such thing as chance." Here "chance" either means "uncaused events" or "the absence of intention" depending on whether the speaker is a determinist or superstitious, respectively.
6. "I would like the chance to speak with you." Here "chance" means a circumstance that lends itself to a purpose, an opportunity.
7. "Quantum chance prevents a simultaneous description of the position and velocity of a particle." Here chance refers to "indeterminism."

In 1, 2, 3, 4b, and 5, "chance" denotes the absence of intention and does not mean "uncaused" or indeterminate but refers to events were that were simply the coincidental intersection of two separate causal chains. For example, there is a chain of causation that got the helicopter over the volcano and there is a chain of causation that resulted in the eruption. The fact that they happened at the same time was not related to either causal chain. That is, the volcano did not erupt *because* the helicopter was above it.

However, the greater the supposed improbability of the coincidence, the more likely superstitious people will attribute it to some divine purpose. They are even more likely to attribute the event to a divine plan if the coincidence has an effect that seems meaningful, if, for instance, the helicopter was being piloted by an evil scientist who didn't believe in purpose. (Novelist Thomas Pynchon noted this kind of chance as "a Hollywood distortion in probability.") So in 3, if the speaker says the volcano blew "by chance" with arched brow and knowing smile, he may mean "by the force of justice."

"Chance" in 4a is itself a cause in an extraordinary sense, a sense outside of normal causality. In 4b "chance" could also refer to a chance likeness between the numbers chosen and drawn, if we understand that in some sense they are not the "same" numbers, for the lottery player picked them for idiosyncratic reasons and the machine "picked" them driven by the laws of physics; in 6, it is part of a purpose. In 7 "chance" refers to indeterminacy or the fact that matter or thingness is not precisely describable or stable. The state of matter is always probabilistic.

At the very least I can say that "chance" is a tricky term. None of the above definitions alone quite suit me and yet all of them, even those that contradict each other, contribute to my understanding of chance and *telos*. It may be useful to describe the working of teleological chance without using the term "chance" at all but by using more specific terms.

Purpose, I argue, involves the *spontaneous* emergence of organization within *stochastically* interacting systems, that is, systems in which the individual parts interact *randomly*. Let me try to clarify what I mean by these various but related terms.

I define *random* as having no discernible repetitive pattern, nor next measurement that can be predicted. This is so probably because the causal factors involved in creating the observed random pattern of behaviors are coming from various unrelated sources, which are constantly changing. Imagine a group of molecules in thermal motion, bouncing around in *unpredictable* ways or randomly. Each molecule is being buffeted by a different molecule, on its own unique trajectory, every few seconds. So if we keep watch on one molecule, it will move in a random way.

Perhaps *stochastic* is better defined by example. I will use a very, very simple system, a chemical reaction, to illustrate how teleological behavior can begin to emerge in a stochastic process. The reader may think I am obliterating the meaning of purposeful human behavior by comparing it to something so mundane. But I want the reader to keep in mind that humans are made up of countless nested, interacting stochastic systems. As Juarrero stresses, purposeful behavior results from:

> *the progressive 'internalization of regulatory processes' that marks the evolution from the proto-autonomy of [self-organizing chemical reactions] ... to the strong autonomy present in [organisms] ..., and finally to that displayed in the exercise of human free will.* ("Top-Down" 85)

Out of the simple comes the complex. And, arguably, this is a more beautiful and profound way of looking at purposeful behavior than to suppose that it must come from something equally mysterious like a fully specified plan.

In the 1950s, scientists Boris Belousov and Anatol Zhabotinsky provided one of the earliest experimental demonstrations of self-organizing chemical reactions. They poured a thin layer of four different kinds of chemical into a Petri dish. The mixture did not remain homogeneously distributed. Instead, it self-organized into concentric and spiral-wave patterns.

How did this occur? As the molecules move about randomly intersecting with this one and that, certain molecules may react if, say, they happen to be "complementary" in some way. I will try to explain what is meant by complementary by an example of pattern formation using typographical symbols that are constrained in the ways they can interact by very simple rules. Belousov and Zhabotinsky's experiment is more complicated (involving four elements), but the basic idea of this kind of self-organization can be illustrated with a simpler example of the formation of spot patterns instead of spirals or targets.

This illustration may seem a bit dull, but don't be tempted to skim. Throughout this book, I will refer back to this illustration as an example a pattern that grows by "chance" nearness and "chance" likeness.

Let's say we have only two elements, which we will label molecule type ╙ and molecule type ╥. They can be turned in any direction, *e.g.,* ╙ can also be placed as ╖ and ╥ can be placed as ╨. We will say that ╙s don't interact with each other; which keeps them stable. The same goes with ╥s.

1. ╙ + ╙ = ╙╙
2. ╥ + ╥ = ╥╥
3. ╖ + ╖ = ╖ ╖
4. ╨ + ╨ = ╨╨.

╥ s and ╙s combined in certain other orientations also result in no change:

5. ╙ + ╥ = ╙╥
6. ╖ + ╨ = ╖╨.

Hence, in the combining operation notated by the "+" sign, found in [1] through [6], there are no changes in arrangements.

But when ╥s and ╙s meet and combine in other ways, they can interact and thereby undergo change, *i.e.,* a ╥ can turn into an ╙ and vice versa. ╥ and ╙can react together and transform each other if they are oriented this way:

7. ╙ + ╨ = ╙╙

8. ╥ + ╖ = ╖╖

9. ╖ + ╥ = ╥╥

10. ╨ + ╙ = ╨╨.

Note that in [7] through [10] whether a ╨ now gets transformed into an ╙ [7], or ╥ now gets transformed into an ╖ [8] depends on the orientations they possess when they meet, that is, on whether the *open horizontal part* of the ╨ or ╥ meets with the open horizontal part of the ╙ or the ╖, respectively. And, whether or not the ╖ gets transformed into a ╥ [9] or the ╙ gets transformed into a ╨ [10], depends on whether or not the *closed vertical side* of the ╖ or ╙ connects with the *open end* of the ╥ or ╨, respectively. These are the *simple rules that limit* interactions.

Because the molecules are always in thermal motion in the dish, the way they happen to meet up is random, a coincidental intersection. Statistically speaking, the production of new ╙s or new ╥s is equally likely. One might think that together these reaction scenarios would tend to average out so that any preponderance of one or the other would "wash out," consequently, maintaining the mixture in a steady homogenous state composed of a random mix.

But the point of the thought experiment is to show how this may not happen. Instead differentiation occurs. Although the mixture can start out with a fairly random distribution, a disproportionate-sized clump of, say, ╙s may happen to form in one area since there is nothing to prevent chance clumping. Randomness is not perfectly non-repetitive but allows for all sorts of possible combinations, groups, and clumpings. No ╥s will be produced in an ╙ clump because a ╥ is required to produce more ╥s. As a result, even more ╙s may be produced at the edges of the clump when ╙s happen to come in contact with ╨s in the appropriate orientation, and the ╙ clump grows. Although, at first impression, it may seem that the probability of large clumps forming is very low, because a clump is self-reinforcing and self-stabilizing, its probability of growing, once it gets started, increases. Remember that when ╚s on the edge of the clump come in contact with ╥s, both types remain unchanged.

Eventually, some ╖ + ╥ combinations producing ╥s are bound to occur. The boundaries of the ╚ clump are now defined. What one sees looking down at the Petri dish are fairly regularly distributed spot patterns, some of ╚ clumps and some of ╥ clumps, floating in random mixture of both ╚s and ╥s turned in various ways. If ╚ = black and ╥ = white, black and white spots will appear on a gray background, as on the coats of some dogs. The emergence of a spot pattern is a regularity that wasn't there before and grows on its own, without, say, a genetic code specifically *for* spot patterns. It may be that genes simply "code for" the ability of molecules to interact in *limited* ways, acting as constraints on the system, resulting in general types of patterns at larger scales.

In more complex systems, the patterns can be more complex. Butterfly wing patterns, for example, are formed in similar self-organizing ways, that is, genes do not *specify* the patterns, only the kinds of chemical molecules that are available and to some extent in what circumstances. The process of self-organization makes systems of self-maintaining regularities or entities out of homogenous mixtures. These regularities then might turn out to be useful, as "eye spots" on butterfly wings are, catching the eye of natural selection (see Alexander, "Neutral Evolution and Aesthetics" and "Nabokov, Teleology and Insect Mimicry").

This thought experiment illustrates the concept of "chance" nearness and "chance" likeness. ╚s and ╩s happen to be *near* each other. They don't *have* to be near each other but there is nothing that prevents them from being near each other. They happen to be *like* each other. That is, a ╩ in the right orientation and position is like an ╚. A ╩ does not have inherent ╚ qualities, but these qualities do get expressed when it comes near an ╚. The relations of contiguity and similarity form a system, a whole, a clump of *like* parts, all ╚s. The clump produces more parts like its own when it encounters ╩s, and therefore, it is self-maintaining, and therefore, it is a "system."

In my view, it is self-reinforcement/self-maintenance that is the first, primal step of teleological phenomena. An entity, in this case a system of like parts ╚s, is self-generated and then sustains itself. Part of the reason why the production of like parts occurs as often as it does (more often than would be "probable" if they were not in and of a system) is because of both differentiations in the system, as well as its boundaries, limitations or constraints.

The production of like parts is a sign of the system's existence as a whole and its constraints. Let's suppose that for some reason you can't see the entire spot of 𝕃s. You happen only to be able to see a small sample of molecules right at the edge of an 𝕃 spot. You notice that more 𝕃s are being created that you would have predicted, and this leads you to infer that there is a big clump of 𝕃s nearby that is affecting the probability of 𝕃 production. That is, you think there must be a "system" forming that is constraining its parts. Thus by means of a *sign* of a whole you have been able to infer the existence of that whole. We can compare also the earlier example of Daisyworld, in which the relative growth of black or white daisies is a sign of stable temperature.

Spontaneous in the sense used here means intrinsically caused, *i.e.,* not the result of an imposition of regularity onto the system from outside it. The molecules in this example form patterns all on their own without any help from a director or prespecified codified instructions. They are the kinds of patterns that *inevitably* form in these chemical mixtures (and don't need natural selection to help create them). We may think of this example as comparable to *autocatalysis* in which the parts of a system contribute to the process by which they are further produced. Please note, though, that "spontaneous" does not mean uncaused.[4]

Random, spontaneous, and stochastic can be used to describe the dynamics behind other patterns besides those made by molecules in chemical systems. Individual trades in the stock market interact relatively randomly, and are not directly synchronized or organized by a central agency. Collectively and spontaneously, trades establish a coherent notion of value. Birds interact stochastically and flocking occurs spontaneously. As Brian Goodwin notes,

> *chemical reactions, aggregating slime mold amoebas, heart cells, neurons, and ants in a colony ... all show similar types of dynamical activity—rhythms,*

4. As Bruce Clarke and Mark Hansen note, cyberneticist Heinz von Foerster was among the first modern thinkers to note that physical difference of materials in Brownian motion may result in the parts falling, by chance, into order, not disorder. In "On Self-Organizing Systems and their Environments" (1960), he asks his readers to imagine magnetic cubes being tossed around until the opposite poles of each connect. The result would not be dissimilar in appearance, von Foerster notes, to abstract art.

waves that propagate in concentric circles or spirals ... The important properties of these complex systems are found less in what they are made of than in the way the parts are related to the whole—their relational order... (23)

Reflecting on the relevance of spontaneous organization in the biological sciences, Evelyn Fox Keller observes,

What was appealing about this view was that it offered a way out of the infinite regress into which thinking about the development of biological structure so often falls. That is, it did not presuppose the existence of a prior pattern, or difference, out of which the observed structure could form. Instead, it offered a mechanism for self-organization in which structure could emerge spontaneously from homogeneity. (*Reflections* 150-151)

My ╥ and ╙ system is an abstraction. There is no such thing as a type ╙ molecule or a type ╥ one. It is not a *real* chemical reaction. But it could represent some of the ways in which orientation and shape might affect chemical reactions or the formation of macromolecules. Complexity workers Terrence Deacon and Jeremy Sherman (2008) have also written about how "morphodynamics"—the way shapes relate—may contribute to the emergence of teleological phenomena. They describe a form that is created by bits first forming triangular pieces that bond together easily because of their complementary shapes (with the help of a catalyst). These pieces happen to come together in larger shapes, also because complementary shapes make bonding more likely than with less regular shapes. A tube-like form self-organizes from this process. Inside the tube some of the original unformed bits are caught, so when the tube happens to break, the raw materials are there together in roughly the right amounts such that they are likely to reform into a tube again.

Deacon and Sherman claim that this may be the way in which proto-life, (*i.e.* self-reproducing, self-maintaining forms) emerged from non-life. Such ideas may be helpful in understanding how molecules with formal properties like genes might have first arisen. So, while my ╥ and ╙ system is simplistic, and Deacon and Sherman's model only a little more complex, such models may be applied to many interesting and indeed, what may turn out to be very profound processes.

I also want to note that it is not an accident that I chose shapes that look like the letters T and L. T and L have very different uses in language; they denote very different sounds *and aren't usually confused with each other.* It is a surprising coincidence that an up-side-down T might function as an L or vice versa. And likewise I want to convey the idea that in this simple system it is surprising that an ╙ might function as a ╥. One would not have predicted it if one had previously only studied Ts or Ls in isolation. The context in which the ╨╙s find themselves defines what they are and how they can function. In this way, a ╨╙ is a pun, that is, its ╥ meaning is changed by context, of its being up-side-down near an ╙.

In this example I have defined "the ordering tendencies of chance" in a way that incorporates the first six different meanings of chance given above. (Since the ╥ and ╙ example occurs above the quantum level, the concept of chance in 7 is not relevant here. I will address that concept in the next chapter.[5]) I noted the relative improbability of chance events. I mentioned coincidental intersections of *coincidentally*-like (not *inherently*-like) parts that eventually result in self-organization. I noted how the effects of chance events were spontaneously caused. Most importantly, I noted how the chance events served a purpose of forming a *self*-organized system, a distinctive self-maintaining entity, or a proto-self.

The emergence of a self/system is, of course, essential to purposeful action because purposes are defined in relation to the selves/systems that they serve.

ARTISTIC SELFHOOD

Now imagine multiple selves/systems nested within larger systems, interacting with or nested within even more systems. Molecules constrained in cells, cells in organs, organs in bodies. Each of these systems is linked to each other in ways fixed by natural selection. The deeper the system is nested, the fewer degrees of freedom its parts have (relative to what they would have if they were not in a complex system). However, the larger whole is highly adaptable because it is made up of so many sub-systems that can re-organize and adapt to new situations as needed. This is,

5. Suffice it here to say that "thingness" in 7 may be a primitive kind of emergence, not complex emergence, a bit like temperature that emerges from a collection of molecules whose various positions and velocities are summed up to give one overall reading on a thermometer. This distinction between different types of emergence is sometimes noted as "weak" and "strong" emergence.

very roughly, how we can segue from self-organizing chemical reactions to self-organized human selfhood. The self-organized chemical reactions going on in your body are not so different from those going on in Petri dishes, but in a body their holistic effects are usefully harnessed by the larger, interrelated systems.

Artistic creativity is also a self-organizing process. Ideas might interact randomly much like the ╙ and ╥ symbols do. Even more freedom and more chance may be possible here, as the parts are abstractions, not actual molecules, and so are more flexible. An idea in an unusual orientation may connect with another idea, if they happen to have a surprising, previously undiscovered likeness in that new context. "Life's nonsense pierces us with strange relation," observes poet Wallace Stevens in 1947 in "Notes Toward a Supreme Fiction" (329-52). Constraining the random interactions are the artist's habits of mind, which determine what goes with what, what is next to what, what is like what.

An artist's habits may follow his or her own idiosyncratic logic, but it is a logic with rules, as Samuel Taylor Coleridge (1772-1834) observes in the famous passage quoted at the start of this chapter. Coleridge's "fugitive" cause is precisely the sort of cause that can make an up-side-down T like an L. Fugitive causes cause coherent wholes to emerge; they make thoughts and create dreams.

As neuroscientist Gerald Edelman stresses, minds are nothing like computers. Thoughts are not inscribed, imprinted, or encoded in any actual physical structure in the brain. Instead they manifest as fluid patterns (like those in our Petri dish) of active neurons and neuronal groups. With imaging technologies, we can see thoughts occurring as electro-chemical patterns that have identities that persist through time in certain areas. It is difficult to know if thoughts take shapes like the spot patterns or spiral waves that form in chemical reactions or other self-organizing systems. Neurons are not all connected side-by-side. A neuron may have a very long dendrite reaching all the way to a neuron in a distant area of the brain. But thoughts, electro-chemical patterns, very likely do have general shapes, very complex ones that to our eyes, through the frame of an fMRI, appear fairly messy.

Neuronal connections can be strengthened the more the neurons interact, and the more they interact the more likely the activation of the one will activate another other. Unused connections may eventually die away. The connections that persist are tenuous, not like actual physical "wires" connecting neuron to neuron, group to group. The connections are more a kind of familiarity than a physical tie. Neurologist Donald Hebb (1904-1985) made the famous pronouncement, "neurons that fire together wire together," but I would that he had said instead, neurons that fire together develop a greater tendency to fire together (though I realize that's not as catchy). The connections can be made, and the physical structure of the brain actually changed by thought, but not so much with a material kind of linkage, but by means of an increased (physically defined) propensity to interact. Those neurons near each other become more like each other as they develop similar behavior together.

Neuronal activations that happen to be near each other in time or space tend to reinforce each other and reduce variety thereby. Just so the upside down ╦ is like an ╚, and if they are in close proximity with one another, they will produce more ╚s and less variety. A neuron or group of neurons is also like a bird in a flock that is constrained by its nearest neighbors. Although it could fly in any direction, it is more likely to fly the way its nearest neighbors are going, and since the same is true of all the neighbors, regularity spreads. We experience this regularity as a more or less coherent thought. This kind of regularity and structure in thoughts emerges from chance associations of near/like patterns (indexical/iconic), and so is semiotically determined as well as materially determined.

In the days when poetry was still largely an oral tradition, poets remembered their lines by reciting each stanza in a different room. When they walked into the next room, they would be reminded of the lines of the next stanza. After a while they did not have to walk the actual rooms, but could just imagine themselves walking from room to room and remembering the lines as they went. In fact, the term "stanza" means "room." We have long known about the odd way our brains work and recall figures and facts using arbitrary associations. Some claim that if you take notes with a certain kind of pen you might do better on an exam if you use that same pen, for its familiarity, the particular flow of its ink and strength of its line, may help you remember the material. If this is true, it isn't magic or luck. It's self-organization. With such knowledge, we show a good intuitive sense of the way memories are "stored," that is, not in a place, but by association, often arbitrary associations.

Sometimes what's near, or *contiguous*, in your mind, to one idea may not be necessarily connected to it, in the way that, say, smoke is necessarily connected to combustion. Let's say the smell of garlic always reminds you of your uncle. This is so because your experience of your uncle was *usually* or often, but not always followed in time by a whiff of garlic. It may be that the garlic pattern often activates the uncle pattern in your mind. Contiguity is related to what in semiotics is called an "index." The smell of garlic points your thought to your uncle, or your memory of him.

At other times, thoughts having some coincidental *similarity* can get called up together. A gnarled twisted old oak branch always reminds you of your aunt's arthritic hands. Similarity is related to what in semiotics is called an "icon."

Connections don't necessarily have to be logical or rational, although most might be, since most of our experiences are experiences of causal relationships: a flame always makes us think "hot." However, there is no reason to suppose that logical ideas and correct knowledge of the world form in ways differently than "poetic" memories and perceptions do. The former are simply more strongly reinforced because they don't have merely contiguous (near) or merely similar (like) connections, but inherent connections.[6]

STOCHASTIC RESONANCE

In neuronal dynamics, creativity can occur when associations, having nothing essential to do with the particular habit of thought that is presently being entertained, are by chance reinforced by other irrelevant stimuli/associations. In the brain, distinctions may not be made between *actual stimuli* and strong *associations* that have formed connections over time. Numerous contingencies associated with previous manifestations of a thought often appear, *e.g.*, one suddenly recalls what one was wearing when last one smelled that scent. A neuron or neuronal group may have strong connections with other neurons and/or groups, and these connected groups may not be relevant to the stimuli that the first group is responding to, but the other groups may become activated all the same. Most irrelevancies must be dampened away or we could never think straight, but some coincidental similarities and contiguities may enhance a few irrelevancies and make them

6. C. S. Peirce wrote about indices and icons, but he did not specifically note what I call "accidental" indices and icons, whose connections are not inherent. The idea of "accidental" connections is necessary to explain how radically new things emerge and become teleological. See Alexander, "The Poetics of Purpose."

come to the fore in the process know as *stochastic resonance* (SR). Let's suppose that you are looking at a flower in very dim light and you cannot tell what color it is. But you have an unconscious memory of seeing that type of flower in full daylight, and that one was blue. You may *actually see* blue in the dim light. In the parlance of SR, the blue flower is a "signal" and your various memories are "noise," which in this case boost the perception of the "weak signal." Or maybe you have never seen that kind of flower (signal) before, but there is a radio on in the next room (noise), just barely audible, which is playing *Blue Danube*. Again you may *actually see* blue in the dim light. In this way, a stochastic resonance may aid your perception. Added "noise" can, if it resonates just right, increase "signal" strength.

The concept of "stochastic resonance" was introduced by scientists in an effort to explain how the Earth's climate responded to certain negligible periodic effects in the solar system, effects that could not be strong enough to make the difference they seemed to be making. Other planets pull on the Earth, making its orbit around the sun eccentric, and so potentially making the Earth very slightly cooler or warmer at times. They found that environmental noise (the unrelated effects of vegetation or atmospheric content, say) that is coincidentally similar to those solar system effects (cooling or warming, say) might strengthen the effectiveness of those perturbing other bodies, resulting in *significant* periodic climatic change on Earth that is consistent with astronomical periodic differences. Since these discoveries were made, examples of stochastic resonance have been identified in numerous other nonlinear dynamical systems, especially biological systems.

We might think of stochastic resonance as synchronicity with real teeth. Like synchronicity, stochastic resonances involve *meaningful* coincidences. But the theory of SR offers a scientific explanation for their effectiveness and origins.

Stochastic resonances can help an entity go on along in its more or less predictable direction. If so, they are *confirming*. Or they may take an entity in a surprising new direction. If so, they are *novelty producing*. Stochastic resonances might make an ecosystem suddenly switch to a new *more adaptive* regime of behavior by allowing it to respond to a previously undetected signal (*i.e.* useful pattern in the environment). Stochastic resonance is, I argue, a type of interpretive response of a complex system to noise *as* signal. The noise isn't *really* signal: the noise is not actually related to whatever pattern might be part of a entity's repertoire of behaviors, which are modeled on the environment; the noise is just similar to it, and so responding to noise as if it were signal may be considered an interpretation. And so it is conceivable that an ecosystem, like Gaia, can make an *interpretation,* if you define it in this way. An inanimate (much less complex) system like a chemical reaction can also make a proto-interpretation. For instance, if a ╩ gets close to a whole "system" of ╚s, the system (one of the ╚s) will respond to it by transforming ╩ (noise) into ╚ (signal), the physical structure that the system needs to grow and continue. This can't be anthropomorphosis of inanimate physical processes if our own brains (groups of neurons) make interpretations in more or less this same way, responding to noise in ways that strengthen patterns. As Jesper Hoffmeyer notes, "The mental system of humans has grown from nature through an evolutionary process, and we must expect to find phenomena in nature that remind us of humanity..." (*Biosemiotics* 6). Although the concept of stochastic resonance has some very specific technical applications in science, we can think of it here more simply as *meaningful luck*. This is a kind of luck that isn't pure coincidence, for the constraints of the system make the meaningful occurrences more likely than not.

Literary theorist Wai Chee Dimock has written about how stochastic resonances in literature can help us make insightful interpretations of literature, by strengthening a signal, or meaning, which may be hidden.[7] At the time Herman Melville wrote "Billy Budd," a short story about a tragic sailor, the term "gay" did not refer to homosexuality. Melville describes Billy as "gay," and although there are other details that might possibly have alerted some of his contemporaries to a homosexual theme in the story, they aren't especially notable. To us however, the coincidence that the term "gay" now has come to refer to homosexuality makes those themes leap off the page.

Of course a stochastic resonance could be wrong and very misleading too. There are many, many bad interpretations of literature based on spurious analogies, noise that boosts the wrong signal. So, too, in everyday perception. You may have seen a type of flower in purple previously, and if you later see a similar blue flower in dim light, you may actually and incorrectly perceive purple. What determine a good versus bad interpretation are the number of other corroborating analogies that support the same general reading. A successful author provides a number of recurring patterns such that the author's self-organized meaning is more or less recreated in the reader.

7. Dimock received a Dactyl Foundation award for her essay, "A Theory of Resonance," in 1998.

In other cases, stochastic resonances may help you have a different kind of reaction or think a new thought. Let's suppose you're thinking of the chemical reaction described above and the way a slight irregularity in the homogenous mixture (all ╙s, no ╥s in a small area) can cause further differentiation (a growing spot of all ╙s) spontaneously. Your thinking of how ╙s get divided from ╥s makes you recall the Biblical lines, "God divided the light from the darkness," and this make you hypothesize that creation by differentiation may indeed have been the way in which the universe was formed, and, as you happen to be a cosmologist, you begin to apply this insight to a new theory. When there is stochastic resonance, there may be sudden insight, a new idea, an epiphany or a spontaneous decision. Inanimate systems that are complex and have feedback, like the ╥ and ╙ mixture, can involve stochastic resonances, likenesses that are idea-like, spontaneously forming systems that further reinforce those "ideas." *All such acts, in nature or in our minds, are quintessentially purposeful acts.*

This is not just "luck," but interpreted luck that sustains or furthers the entity making the interpretation. Interpretation is always a subjective response, not a mechanical reaction. A clod of dirt can't keep itself together by selectively interacting with rain, wind and temperature; it is not a complex system and so reacts mechanically to physical forces. An organism *can* keep itself together by selectively accepting what is good for it and rejecting what is not. And so an organism is purposeful, while a clod is not (*pace* Blake).

ESTRANGEMENT

As this theory of purposeful behavior is very much reliant upon self-causation, it is also a theory of creativity or original behavior. It is equally a study of what makes art artful. In 1914-30 in Russia there was a movement that attempted to describe what is artful about poetic language. The short summary is: they found artists tend to *estrange* language. But if this is taken too literally, you might have artistic responses in the form of intentional disruptions of language, making the sense unclear in a confusing and boring way rather than making a new sense.

I like the movement's principal thinker, Viktor Shklovsky (1893-1984). He writes, "Countless stories are, at bottom, extended puns" (53). Although he overstates his case and probably only means the stories he likes are extended puns, he is right, I think, in noting here that *estrangement must make a new sense, not merely destroy the former one.* Above I mentioned that a ╧ is a pun, that is, its meaning is changed by a changed context. A ╧ is estranged by being in the up-side-down position, but just turning it over doesn't make it interesting or artful. It's the pairing it up in the right way with an ╚, which surprisingly results in a reaction of more ╚s and eventually spot patterns, that makes the pun on ╧ artful.

Literariness, *poesis*, and creation, argues Shklovsky, depend upon rule distortion whereby new functions (new meanings) are found for pre-existing patterns. He claims "art distorts ... a correspondence in accordance with its own laws." Shklovsky shows, as I have tried to do with my example of the simple self-organizing system, that these laws are formed by a combination of chance and constraints. They emerge suddenly in poetic resemblances and chance patterns. "Not surprisingly," writes Shklovsky, "the author himself may have a hard time recognizing his own work" (171). That is, the author may be surprised to find an intention forming in his work of which he had not consciously been aware.

According to Shklovsky, this is what the much-admired Russian writer Alexander Blok experienced in 1918 in the course of composing *The Twelve,* a long poem about twelve Bolshevik soldiers (like Christ's twelve apostles) marching in a raging blizzard.

> *Blok...began* The Twelve *with street talk and racy doggerel and ended up with the figure of Christ.... The "Christ" finale serves as a kind of closing epigraph [gnome], in which the riddle of the poem is unexpectedly solved.... "I don't like the ending of* The Twelve *either," said Blok. "I wanted a different ending. After finishing it, I was myself astonished and wondered: Why Christ, after all? ... Yet, the more I looked at it, the more clearly I saw Christ...."*

Shklovsky describes how Blok's Christ emerged out of some random details Blok had by chance included in an initial draft. Blok was describing people walking down a street forming a procession and the wind blowing in their banners.

> *The wind rips the banners. The wind in turn calls forth the flag, and the flag, finally, calls forth someone enormous bearing a certain relationship to it. It is precisely at this point that Christ appears on the scene. ...he was called forth by the compositional pattern of these images.* (171-72)

Blok's experience is reminiscent of my own experience of my intentions emerging in the self-organizing process of writing. The experience is common among artists. Following Michael Polanyi, biosemiotician and cultural theorist Wendy Wheeler[8] similarly notes that creativity:

> *lies in being able to "disattend from" the logic and rules of grammar, and other rule and grammar-like governed forms of semiosis, in order to "attend to" the perfusion, and disorderly profusion, of many signs which aren't supposed to count as legitimate when semiosis is thought about only in linguistic rule-bound terms. This "attending to" is both conscious and unconscious—a kind of free-floating attentiveness...which poets have long described as waiting for the muse to descend.* (*Whole Creature* 146)

Understanding the behind-the-scenes processes of creating is necessary to seeing art as a self-organizing system. It may not be possible to look at a painting or to read a story or poem and realize that it was formed in this loose, organic way. The final product may seem quite coherent and put together in a way that, as most might say, seems completely "intentional." Well, I would agree, but we might not have quite the same definition of "intentional" in mind. People often use the term "intentional" to mean pre-planned, as if art could unfold from a seed of an idea that already contains fully complete instructions. Absurd! If I note something looks pre-planned (and I see that the artist is not particularly sophisticated or able to redevelop artistically a pre-planned design), I might say it looks formulaic or mechanistic. I reserve the term "intentional" for those artworks that develop in the *organic* way that I have described.

I want to note here that some pre-planning a work is not a bad thing in art. We ought to start with a language, a system, a pattern as a kind of raw material. Our product will be very different from the plan, which can only be vague and shadowy because details can only get worked out in the self-organizing process of creation. It's what we do with that vague pattern that makes the work original or artful or not. If we do not start out with any rules (differences that are constraints), then the best we can hope for is a random mixture, something like soup (*á la* Jackson Pollock, 1912-1956). If we start with soup, which contains random mixtures of different things (having simple patterns or formal qualities), and repeat some sort of rule over and over, we

8. Wendy Wheeler received an award in 2009 from Dactyl Foundation for her essay "Creative Evolution: A Theory of Cultural Sustainability."

might come up with something as interesting and artful as spiral patterns in a Petri dish (like the "better" Pollocks if you stare at them long enough).[9] But we're capable of so much more than that. We have the very plastic mediums of language, convention, and tradition to work with. We have much more interesting raw materials than ╙s and ╥s.

It seems to me that this world is complicated and confusing enough without working hard to estrange it. Estrangement is all around us: things randomly colliding, things we mistakenly see in up-side-down ways, things resonating stochastically. The artistic mind is the mind that can't help but see stochastic resonances. The artist's role may not be to estrange the world so much as to notice and make new sense of the naturally occurring estrangements. The artistic mind is a keen observer. It can detect the slightest hint of a pattern, even or especially chance ones. This the artist has in common with the madman. What the artist can do that the madman cannot is keep his mind from running off in too many directions, by constraining the variety into a coherent whole. Being able to do so indicates selfhood.

New York-based artist Neil Grayson has helped me apply my understanding of the process of self-organizing intentionality to the process of creating visual art. In the next chapter, I describe how he and I started the Dactyl Foundation to help artists who identify with the artistic process in this way and to promote their work. This book, one might say, is one long call for papers and projects related to the subject of teleology and art, art and science, teleology and chance.

In English, the word "dactyl" is used most often in poetics and describes a type of meter with one long stress and two short stresses. The word "gallery" is dactylic. Dactylic stress to me sounds energetic and fitful. The word comes from the Greek for "digit": your fingers have one long joint and two short ones. "Dactyl" is sometimes also used in the arts to describe a relationship between the old and the new, between the this and the that, which one might think would be unbalanced or awkward, but which actually hangs together with surprising coherence. Thus, the Dactyl Foundation takes its name, supporting aesthetic approaches that understand the emergent ordering tendencies of

9. Physicist Richard Taylor studied Pollock's paintings and found, not surprisingly, that their structure could be described in terms of fractal patterns, that is, patterns that repeat the same or similar structure at larger and larger scales. Such patterns tend to emerge naturally when the similar rules are applied over and over.

chance and how things near each other and/or like each other can interact to form organization that is both regular and surprising.

Because my expertise lies in literature not visual art, Grayson has helped me with the parts of this book that relate to the teleology of painting. He has shown me how chance is employed in the artistic process to reveal unconscious tendencies and themes. I can illustrate this with the story of how he did one of his pieces called *Man Being Led Away* (1994). It's helpful to have access to the description of the process that goes into making a painting if one wants to understand the intentionality of the work. I'm lucky to have Grayson as well as Dactyl-supported artists who are able to articulate this complex and largely unconscious process.

As Grayson describes the painting in literal terms, "One figure stands gaping at something. His fist is clenched. A second figure is trying to get him to turn away." But there is also a strong theme suggested by the scene. Why is the second figure trying to get him to turn away and what has the first figure done, seen or realized? The theme, explains Grayson,

> *happened accidentally. I had originally painted a single figure, but one of his arms didn't seem to belong to him, and I painted in the second figure behind the first, gave that arm to him, then I painted another arm making a fist for the first figure. A few weeks later [after the painting was complete], I was telling a story about my father. I wasn't thinking about the painting at all. My father was one of the first Jews ever to be admitted to West Point. That was 1942. My father was hazed and beaten, especially by one of the sergeants. Finally, my father had had enough. He tried to kill the guy. His friend had to drag him off. When my father saw the sergeant's head bleeding, he couldn't believe what he'd done. He couldn't unclench his fist. He was sent to a mental hospital and given shock treatments, during which his wrist bone broke...*

In the midst of relating the episode at West Point, Grayson connected it to the painting. The painting was the third or fourth in a series of work later called "The Fighter Series," which he had begun almost immediately after the sudden and premature dementia and death of his father. Grayson knew that the burst of creativity was in response to his father's death, but he hadn't understood how exactly and why a "fighter" might be relevant to his father, a Jewish lawyer. It can be noted too that the man being led away in the painting does not resemble Grayson's father, instead he slightly resembles one of Grayson's friends, who had also recently lost his father suddenly and tragically and who had suffered a mental breakdown as a result. This is a nice example of the crazy cross-referencing and consequent unpredictability of the artistic mind. Art and dreams give us our best evidence that the mind is self-organizing, making use of stochastic resonances. Like Blok, Grayson noted that in retrospect it seemed as if he had painted the two figures for the purpose of illustrating his father's trauma. He observes that "art is the employment of the chaos."

After discovering a theme spontaneously, or through a stochastic resonance, the artist may then try to bring out that theme more clearly. It is in this regard that the creative process begins to resemble slightly the process of natural selection, which edits and culls and makes a particular kind of organization more stable and able to reproduce more true to type. Undoubtedly the realization and understanding of the meaning of the fighter theme affected how Grayson painted the remaining paintings in the series.

An artist's "periods," such as Picasso's "Rose" and "Blue" periods, often reflect, not just a chunk of time, or even a particular palette or subject matter, but an "aboutness" that abstractly represents the artist's feelings and experiences. Consider, *The Tumblers* (1905) or *Harlequin Sitting on a Red Couch* (1905), or *Acrobat with Young Harlequin* (1905), which belong to Picasso's Rose period and feature circus performers. There is a conventional meaning or theme here, which has to do with a stereotype for such professionals and the way they lived, but there are also signs of Picasso's more personal experiences and sympathies with these people in the postures and expressions he tended to paint.

Despite his apparent mastery of the figure in Picasso's student work, for example *First Communion* (1896), one cannot say with confidence that he was always capable of representing figures realistically if he wanted to. His figures are often flat and angular, a bit awkward, possibly because his skills dictated such, but if this is true, it's also true that this would-be failing of Picasso's is also his greatest strength, for these saltimbanques and acrobats are only awkwardly fitted into society. As an artist Picasso clearly sympathized. So then did Picasso *mean* to express his sympathies? Or was he simply incapable of painting otherwise? And what is the difference? At one point did he notice this awkwardness in his skill, in himself vis-à-vis society and in the circus people and embrace it as the series' theme?

Is the artist "responsible" for this meaning if it comes to him by limitations, mistake or chance? I think he is, for the particular kinds of mistakes an artist makes help define who he is. Who he is guides him as a flexible kind of constraint.

In the self-organizing chemical reaction and self-organizing artistic selfhood examples, I use "spontaneity" as a synonym to "purpose," not an antonym. A purposeful action is self-caused. *Causa sui.* It is the inventing of an organization or plan. It is not the absence of plan. So what came first, the idea or the artist? Do the ideas make a man an artist? Or does the artist make the ideas? Chicken and egg questions reveal the poor logic we apply to some of our most difficult questions. In the complexity sciences, we say that complex systems "self-organize," which is to say that they spontaneously form or are self-creating. They create selves. I wish I could merge the word order of this phrase because there isn't a self that exists first and then does the organizing of itself. The organizing and the self-creating are simultaneous. There is no "self" that is separate from the action of organizing. The self is the process; it's not a separate thing. Likewise human selves are not homunculi-like souls controlling our actions, nor do people have essential natures that determine their thoughts. A self is a process of self-organization; it is the whole that is reflected in the parts of the system, which is ever changing and thereby maintaining a dynamic stability.

Chapter 3

DACTYL FOUNDATION

Others there are who, indeed, believe that chance is a cause, but that it is inscrutable to human intelligence, as being a divine thing and full of mystery.

–Aristotle

There are two major arguments against teleology, that of the *indeterminists* who believe that all outcomes are affected somewhat by chance and therefore teleology is disproved, and that of the *determinists* who believe there is no such thing as chance and therefore teleology is disproved. We may also think of these groups as the postmodernists and scientific reductionists, respectively. These two kinds of anti-teleology groups couldn't be less like in their understanding of teleology, and yet they are united against it. Both groups argue that purpose is only imagined by the naïve masses whose minds are enslaved by bias and desire. The fact that many of my friends and colleagues belong to one or the other group helps explain why I am writing this book. I walk a thin line, arguing against one side, while trying hard not to align myself with the other.

Since 1997, Neil Grayson and I have run the arts organization in SoHo, New York City called the Dactyl Foundation. We are dedicated to exploring and encouraging intersections between the arts and sciences. We bring artists, poets, novelists together with scientists and philosophers of science. We organize conferences, put on lectures, readings, and exhibitions, give support to researchers for travel to art-science meetings, and host a weekly open discussion forum that we call, only a little jokily, the Compost Modern Forum.

We try to make what community we can out of people who have a variety of interests in art, and this means working with indeterminists and determinists. If I limited my interactions to people who share my views on teleology, biosemiotics, and complexity science, I'd have an audience of five or six, and one would be my intern. It can be lonely being too original. So we are open to all. What a mixed bag of participants and artists Dactyl has had over the years! Our roster of invited lecturers is positively schizophrenic. We have had John Ashbery and Steven Pinker read and lecture at the same conference. We have had an exquisite small Goya and a big ugly Basquiat on the same wall.

Another little teleological tale is in order here to tell how Neil Grayson came to be cofounder of Dactyl. Grayson had, like I had, tried to be a proper postmodern artist, but even his squiggles would turn out looking like gracefully suggestive lines that would resolve themselves into figures. Grayson is tall and formidable, ostensibly gloomy and serious. However, when you get to know him, you may be surprised to find how much he thinks in puns, sometimes paying more attention to the sounds of words than their senses. This is perhaps related to the fact that, like a number of artists of Russian descent, he has synaesthesia, a cross-modal perception of the world. He connects colors and textures with letters and numbers—for instance the letter A is smooth cornflower blue to him and Q is bumpy mossy green—and this seems to have made him especially good at remembering arbitrary figures and facts through associative mnemonics. He is amused and enchanted by all chance patterns and responds to them with a quick wit. Once he was introduced to a woman who was named Grace Neilson. I will never forget his instant reply, "Meet to nice you."

He says that as a child he spent many hours staring off into the far distance, playing a counting game in his head. Standing motionless in the midst of a chaotic playground, he spent his recess time visualizing a pattern of a certain number of dots and then using that pattern as the first unit of a larger pattern of the same shape. He says he remembers feeling that the cosmos was similarly ordered and that he was a part in a larger whole that he could sense but never know. Today we refer to such patterns as fractals, and scientists have found that in nature patterns often do repeat at larger and larger scales.

Grayson's agile intelligence, some instances of which could be reminiscent of certain thought disorders or autism, might have tended to be too unbridled to produce meaningful artwork were it not for his training and practice in the science of image. He works his canvases over and over again—and it is through the constant reiteration that the patterns cohere. This, I believe, has directed his talents and has allowed him to do meaningful and astonishingly creative work.

As a young teenager, Grayson apprenticed to a classically trained art professor who gave him the busy task of making color charts with thousands of gradually changing hues. But Grayson's special sense of the significance of color made the assignment an infinitely poetic experience, and he took to it with enthusiasm, learning to make extremely fine distinctions between what most people would consider identical colors. I've been to the paint section in a hardware store with him, and I can tell you that he can pick out the exact match of a color from memory, without a color swatch, name or brand. It is an intuitive but seldom-mentioned fact that those who regularly practice making fine distinctions develop more objective perceptions, and even when the distinction is actually below the human ability to consciously perceive it, practiced observers are able to make more accurate guesses than the unpracticed observers.[1]

Grayson, also well-practiced in eye-hand coordination, developed such freakishly good aim that he might have been a marksman instead of an artist. I have seen him throw a penknife to stab a mark on the wall from twenty feet away. When we're hanging shows at Dactyl, while the rest of us have our tape measures and levels out, Grayson just eyes the wall and locates dead center, and he can even take into consideration the slight pitch of the floor. This, in part, may help explain what Angus Fletcher, in "Neil Grayson and 'This Living Hand,'" identifies as "a strong haptic quality," in his paintings "as if we were touching the pictures in seeing them" (3). This effect, claims Grayson, is probably due to his habit of reworking the canvases over and over, eye and hand going back and forth.

As an apprentice, Grayson focused on some very old techniques, experimenting with the way colors react to depth and layering and how colors affect each other and interact with light. Such techniques, perfected in the Renaissance, were eventually given theoretical treatment by Goethe in 1810. Initially dismissed by physicists, and later corroborated by the complexity sciences, Goethe's theories describe the phenomena of the *perception* of color, which is distinct from the optical spectrum observed by Newton. The importance of understanding color in this way, says Grayson, lies in the way it shows the great importance of subjectivity. It shows that colors have meaning insofar as they are what they are in relationship to other colors nearby.

1. According to T. L. Short, Charles Peirce made the first empirical study of this matter, showing how objectivity is possible under conditions in which it seems objectivity would be impossible. See "Measurement and Philosophy."

Already an expert in these sciences at sixteen, Grayson set up his easel at the Metropolitan Museum in New York and copied Rembrandt's 1660 self-portrait. Participating in the process of a work of art is one way to come to understand its intentions. As he put the layers carefully down and discovered things that he had not expected, he had the exciting thought that perhaps Rembrandt had made similar discoveries in his time. He realized he might be reliving, in paint, Rembrandt's experience. By the time he was done he felt that he had actually spoken with the old Dutchman. I believe this is an instance in which the artist's intention was partly communicated to a very patient and sensitive viewer.

Leaving the museum with his finished version, the brooding young Grayson was detained by the guards, and a special curator had to be called in to confirm he wasn't taking the original.

As Grayson matured as an artist, he applied his training and skills to a new form of image-making for the 21st century. He says he is currently struggling to create profound imagery in a "high major key," which is in some sense is the opposite of Rembrandt's warm glow that so easily suggests depth and meaning emerging from the darkness. Grayson feels the need to respond to contemporary visual habits by "hiding significance in plain sight."

Grayson and I met years ago at a proverbial crossroads, at a public place far from our respective homes, and when an upset espresso splashed onto a white page, we simultaneously made comments that revealed the unlikely fact that we both owned the same print of an obscure Victor Hugo. Hugo sometimes used dark coffee as an inky paint. It seemed inevitable that we should do something in art together, and we've been friends ever since.

At Dactyl we tend not to look for artists who rely exclusively on computer technology to create their work because, as Grayson argues, the embodied mind is such an important part of the creative process, and our self-organizing creative tendencies depend upon our beautifully imperfect perceptions and interpretations. Even our limitations and mistakes can help us create art that is intensely meaningful. Science, in the sense of knowledge not technology, can be enabling, not stifling as some contemporary art rebels believe it is. Grayson says the more he learns the more room he finds to make interesting errors.

Grayson does not talk about his training, and I may do him an injustice here by mentioning it. These days, skilled artists are associated with photorealistic figure studies, with paintings of stiff-postured people posing nude in harshly-lit rooms.

Beyond this book, however, Grayson's science and training are his easily-kept secret. In general, contemporary art critics and viewers are not able to notice the difference between work that anyone could have done and work that very few people are capable of doing. Once a well-known art critic, who, it happens, was very enthusiastic about Grayson's work, paid a visit to his studio. The critic went around looking at various pieces making heady comments, showing real appreciation. Then he turned to a painting by Grayson's four-year-old son that happened to be in the studio. Without skipping a beat, the critic went on praising the work, applying exactly the same kind of pretentious phrases to four-year-old's masterpiece as to the father's work. Grayson couldn't keep from laughing as he told the story. Fortunately, Grayson's sense humor is particularly fond of tragicomedy.

In the beginning when we founded Dactyl, Grayson and I were optimistic that we could have a positive effect on the art world. Our first problem was finding artists to show. There simply aren't enough trained artists whose work seems of this age. With very few exceptions, science in art is terribly uncool. Classically trained artists seem to like to paint figures from Greek mythology or, worse, they seem to be destined to end up making head-shop posters dusted with phosphorous. Some artists who claim to be interested in science don't actually use it to create their work. Instead they illustrate science themes; for example, they paint pictures of DNA or fractals or they get "inspired" by quantum mechanics. I say that art is no handmaiden to science, and it is not art's job merely to illustrate scientific concepts.

Consequently, Dactyl has tended to prefer artists whose styles lean more toward neo-expressionism and postmodernism than realism, even if some of these artists are indifferent to the science of art. Here at least there is more creativity and a desire to have contemporary significance.

Ever since around the time C. P. Snow noted the separation of the arts and the sciences, the sciences have become increasingly reductive and hostile to an artistic sense of the world. Artists have become dismissive of scientists. Artists and scientists have grown more and more apart, and this has had as negative an effect on science as it has on art. The separation noted by Snow, however, is really between Great Books Curriculum conservatives and technocrats, or the

Classics versus Engineering, not Art versus Science. If we think of science in it's original meaning as "knowledge," then we find its interests are not so far from art's, which is also concerned with knowledge of the world and especially our impressions of it. Grayson and I haven't given up on our original mission to bring the sciences back into the arts, but we've realized it will be a lot more difficult than we first imagined.

As an arts foundation, Dactyl was necessarily thrown in amongst postmodernists. That's a given. What's unusual about Dactyl is we are also involved with the kind of scientists who are usually, rightly or wrongly, associated with reductionism. We have often co-hosted events, lectures, and conferences with the Center for Inquiry (CFI), a secular organization with an interest in bringing people to science through art. In college, as mentioned earlier, I had joined various secular humanists groups, including CFI, and that relationship continued beyond my college years, despite my growing interest in teleology. CFI members tend to be guided by a school of thinking called Philosophical Naturalism. Also known as "materialists" or "determinists," their celebrity members include Daniel Dennett, Steven Pinker, and Richard Dawkins. They tend to believe in the objectivity of science; they uphold the values of the Enlightenment and Neo-Darwinism; they tend to pooh-pooh any talk about "emergence" and "complexity," believing that workers in the field unscientifically believe in "strange new complexity forces," as Dennett puts it (50). (You, reader, know by now that there are no funny "forces" in my theory, only chance intersections, coincidental similarities, and constraints.)

I bring determinists and indeterminists together at Dactyl Foundation because I see each group has valid points, and I think they ought to take each other more seriously. There is nothing to be gained from preaching to the choir. I hate to think of myself as a moderate, which connotes being wishy-washy and making compromises, but I do hope that the two different groups can see each other's points (as I have). I'm not advocating tolerance for each other, or the peaceful coexistence of mutually exclusive theories. Not at all. I hope they will shake each other up and change from the experience. Unfortunately, when I bring them together at Dactyl, they tend to annihilate each other's argument instantly like matter and anti-matter. But sometimes the meeting can be slightly productive of new thought, and so I continue, full of hope.

What I like about indeterminists: Postmodernists, most of whom hold a world-view I refer to as "indeterminism," have long had an interest in the way science is similar to art. I agree that scientific discoveries can often involve radical creativity, so Dactyl quite often finds itself full postmodernists talking about how conventions can determine scientific practices and inform theories. They are critics of "essentialism," analytic philosophy's confidence in language and logic, Plato, Truth, Beauty, and God. In addition to my agreeing, somewhat, with the analogy between science and art, I also agree, generally, with the "post-human" position that intelligence is distributed, that is, "we" are not just our brains (or the genes that determine our brains). Rather, who we are and how we act are affected by interactions between brain, body, and environment, which includes our immediate social group and our larger cultural group. I also share the belief that human action is inherently unpredictable and that scientific reductionism with regard to human behavior is deeply incorrect (although we can sometimes make *general* predictions about human behavior, we cannot make *precise* predictions).

I also agree that the fundamentally indeterminate nature of the quantum world deserves our attention and compels us to revise the theories of causality that were born of the Enlightenment. We have to give up the illusion that things have inherent qualities and the world is ultimately made of up *stuff,* as we often think of it, *i.e.* static, definable objects that can be precisely measured. There is no *Ding on sich*. And although I do not advocate radical philosophical relativism, I do an intellectual pluralism á la, say, John Dupré, insofar as there are different truths that correspond to different *levels* of description.

On the other hand, I also agree with the determinists that stuff does come into existence above the level of the quantum world, and here it behaves pretty much as Newton thought it did (as long as we ignore possible holistic effects). We might refer to this condition of predictability these days, as inevitability, rather than predeterminism. The former connotes a probabilistic determinism, which is no less escapable than the latter. I agree with the empirical method in principal, and I think science can be progressive. It sometimes makes wrong turns along the way, so the path is crooked, but it does lead us forward, eventually. I agree with the scientific findings that we do not make conscious choices; rather choices are made as electro-chemical activity in the brain, and we become aware of the choices we have made seconds later. Our awareness then feedbacks into ourselves, affecting future bodily choices. I think it's useful to categorize some things and to make generalizations, and I do feel that abstractions can capture some aspects of reality. I believe in the dynamic existence of reality. I am more a Realist, than a Nominalist.

Both groups have helped me, in different ways, abandon religious notions that I had been taught as a child (though I continue to have an emotional fondness for them as poetic stories). Interacting with both groups, instead of favoring only one, helps keep my mind open.

When I say I take both determinists and indeterminists seriously I mean I meet each on his ground and I argue on his terms. If I'm talking to an indeterminist, I assume all of reality is at bottom "indeterminate," (meaning not miraculously a-causal, but simply not specified with regard to position *and* velocity) and I note that determinism emerges therefrom.

If I'm talking to a determinist, I assume, as complexity scientists do, that we live in a deterministic world, and, even if there were no quantum fluctuations (no microphysical indeterminacy to "exaggerate" or to bring up to macrophysical expression), the dynamic interactions of fully deterministic systems can still create new constraints that enable a system to behave unpredictably. I can take either position, without contradiction, for I believe that determinism emerges from indeterminism and I believe that a deterministic world holds many surprises for those who inhabit it.

THE DETERMINISTS

One evening at Dactyl Foundation, we co-hosted an event with the Center for Inquiry, whose organizer Austin Dacey arranged for Steven Pinker to speak. Pinker is considered, amongst postmodernists, to be a genetic reductionist, and he is an outspoken critic of postmodernisms. The audience literally divided itself with determinists on the right and indeterminists on the left. I was hoping Pinker would dare to talk about his criticism of postmodern relativism he had lately published in *The Blank State.* In that book, Pinker, it so happens, declares that something like calendar art is most aesthetically pleasing to humans, objectively so. He claims science can explain why we can objectively prefer some forms of art over others. Beauty is not subjective, he claims.

Some of my postmodern colleagues can't understand why I consort with these reductionists, who to them are so clearly *wrong* about the supposed objectivity of science it's not worth arguing about it. But I confess I like the look of a picture of pretty rolling hills (*i.e.* calendar art) better than a collage with doll parts, text, and acrylic paint. Pinker has a point. However, it's true that I cannot agree that calendar art is "art" in the sense that I like to think of it. A pretty scene is not necessarily teleological, meaningful or poetic. That is, it's not significant of something else: it doesn't *mean* anything. It's just pretty. Nature is purposeful when you look at it from an evolutionary perspective, when you comprehend how it has self-organized and evolved spontaneously. A picture of something pretty in nature doesn't always show this. Art does.

Postmodern collages often seem suggestively meaningful, even if it's just their ugliness that compels us to imagine that there must be some reason for the apparent mess. The effect can be one of provocative eeriness, nostalgia for a lost *telos* that, postmodernists say , never existed. I will discuss this further in the last chapter of this book, in which I explain why I think it's better for an artist to evolve from postmodernism and not retreat to humanism or classicism.

On the evening of Pinker's lecture, I never got to discuss this with either set of colleagues. The hoped-for showdown never took place. Pinker gave a "safe" lecture on grammar that was entertaining to all and avoided all controversy, then he wisely ducked out after taking a couple of questions from CFI members.

THE INDETERMINISTS

On another occasion, I had invited a postmodernist/indeterminist to sit on a panel on teleology with me, knowing full well that he would caricature teleology as theology. I had learned of Arkady Plotnitsky's work in a 1997 book published by Duke University Press called *Mathematics, Science and Postclassical Theory*. For me, Plotnitsky captured very accurately the spirit of the postmodern view of science, and I think of him as one of Derrida's avatars. I use him here as a stand-in for Derrida, whose philosophy is too much of a moving target. Plotnitsky made a limited point on the panel, which I take to be generally indicative of some aspects of postmodernism.

Plotnitsky and I are both members of the Society for Literature, Science and the Arts (SLSA), and Dactyl was hosting their 20th annual conference in 2006. I was a newish member at the time, and Plotnitsky one of the old guard. The organization had been undergoing internal change for some years. Originally composed mainly of those whose task it was to critique science's supposed objectivity and expose science as "just another narrative" about the world, by 2004 the organization was developing more of a membership that included people working in "system sciences," an area from which the complexity sciences emerged. Inspired by the leadership of Bruce "Bruno" Clarke (incidentally former bass player for the original Sha Na Na), SLSA was beginning to become fertile ground out of which new ideas about teleology might grow.

In his talk, Plotnitsky said, "the very ideas of both history and teleology, at least from Hegel [1770-1831] on, [are] essentially linked to how we think about chance." Teleology, he claimed, entails a "classical" notion of chance behind which there is a hidden or unknown necessity. He argued that final cause is predetermined by the end, and therefore the discovery of quantum indeterminacy makes such causality impossible, as quantum or "nonclassical" chance does not assume a "necessity or causality behind" it.

Plotnitsky supposed that teleology and intentionality are only possible in a world that is strictly deterministic. Such an idea would baffle most reductionists, not to mention teleologists. I told Plotnitsky that his view of teleology had been corrupted by Christian appropriations, and that the "end" in teleology is not physically predetermined.

Neither purposeful action nor *telos* can exist in a timeless Christian-deterministic universe where everything has always already occurred. This is the point of difficulty with all religious systems that espouse an omniscient and beneficent god who determines all things from outside of time. Likewise, purposeful action cannot take place in a classical mechanistic universe where every event is essentially the direct result of the sum of what came before. There is no room for a theory of purposeful action in a model (in which there is a direct relationship between a gene, say, its product and its function) any more than there would be in the Christian model (in which all is directly determined by God). In a classical determinist universe, presumably, one could predict how a person would react to new situations, as long as one has sufficient information. Insofar as this strong deterministic view holds, actions as well as natural events, then, would be predetermined, neither, purposeful, creative nor free.

Someone in the audience then opposed Plotnitsky's deterministic model of purposeful action with an equally dubious model based on indeterminacy.

He noted that the "Action Painters" of the 20th century, for example, argued that purposeful action must be freely executed, not inevitably determined by tradition, convention, past experiences, or habit, or otherwise compelled by any external cause. He espoused "spontaneity," and seemed to be defining that term as "a-causality." In his view, a purposeful, creative action is physically uncaused and thus caused by the artist.

Although I was grateful for the "support" against Plotnitsky, I believe the view that "indeterminism" provides for free will is nonsensical, as determinists such as Dennett have argued. Essentially arbitrary and undetermined actions do not seem any more telic than completely predictable actions do. Many determinists have assumed that this is the position I take when I argue for purpose and intentionality. They think I'm arguing for a freedom of action based on an absence of cause somewhere along the line. I do not believe in a-causality. I argue for a *different kind* of causality, in which the effectual factors come from emergent holistic features.

Neither completely determined nor completely undetermined actions constitute intentional behavior. Radical indeterminism and radical freedom of action seem as opposed to intentionality as physical predeterminism and predetermined action are. What both views lack is a conception of how "probabilistic necessity"—a determinism that arises from quantum indeterminacy—and emergent nonlinearity work *together*, providing direction *and* creativity.

Feeling a bit beaten by Plotnitsky, who is my better in publication and reputation, I attempted to explain that final cause is *emergent*. He replied to the effect that I was interpreting teleology in a way completely inconsistent with the way every one else understood it.

I am always a bit worried that I am guilty of too original, or possibly even incorrect, thinking. And a comment like his delivered in public can make me lose my moorings. I blushed and stammered. (I'm a lousy public speaker under stressful conditions.) Postmodernists in the audience rallied, and I started to wonder if it was such a good idea to put myself on a panel with someone who disagreed so strongly with my position. Sometimes I put too much faith in the power of discussion for resolving conflicts.

Then, happily, a few hands went up in my defense. Someone mentioned how teleology is based on circular causality (as I've argued) not linear causality.

Someone else mentioned Henri Bergson as an advocate of emergent teleology as described in *Creative Evolution*. Another person went so far as to make a comparison between Bergson and Dennett's argument for the existence of free will in *Freedom Evolves*. Postmodernists and reductionists, on this occasion at least, were talking! And the ground they met upon was complexity science.

In the last few years, I have seen the number of people interested in teleology increase, largely in the burgeoning field of biosemiotics. These people usually have very eclectic interests, as I do, and have found surprising ways of reconciling positions that seemed irreconcilable before.

DACTYL ARTISTS

Meanwhile in the visual arts there has been an increase in the number of artists interested in painterly representation. Dactyl Foundation currently shows a number of representational artists, notably Judy Glantzman, Yelena Yemchuck and Sage Vaughn.

Judy Glantzman is an accomplished artist with a thirty-year exhibition history, and she has been with Dactyl Foundation since its earliest days. She is one of the few artists to show very successfully during the 80s and the conceptual 90s *and* to continue to flourish today. Like Grayson, she has learned to trust her mistakes as sources of growth and exploration, treating them like the wise and riddling sages they are. Her subjects are and have always been introspective and psychological—self-portraits, painted as in a dream. With each work, she says she finds a person that *she didn't know she knew*, that, in coming to know, *she helps make*. Her many layered images capture many moments, but, significantly, only as they are *about to happen*. She has often noted that just *before* the instant of realization, she is already sensitive to it. It seems that being an artist requires just this degree of bodily prescience.

Yelena Yemchuk has expressed a similar feeling with regard to the creation of her work: the eureka moment always feeling strangely like déjà vu. In some sense, the artist always already knows what she is just discovering for the first time.

Yelena first began showing photography at Dactyl. Her black and white work, widely known on Smashing Pumpkin covers, features sad and odd circus performers and would-be silent film actors. Gradually, she developed a talent for painting, and turned to colorful imagery from her childhood in the Ukraine, with dark forests, eerie green swamps and primitive huts, life-loving gypsies and odd little churches. Her work, familiar and strange at the same time, seems highly suggestive of some narrative line from a fable, but you can't quite figure out which.

At first glance, Sage Vaughn's work may seem to depict pretty, bucolic suburban life. On second glance, one notes that the sparrows and pigeons are tattooed with gang affiliations. Telephone poles have replaced totem poles and smoke stacks, steeples in their sacred significance. In Sage's world, wild animals have become partially domesticated and children have gone feral.

All three of these painters have in common a sense of teleological meaning in their work that is very unlike what any postmodernist would call "purposeful."

Chapter 4

SOMETHING OUT OF NOTHING

Action on the move creates its own route; creates to a very great extent the conditions under which is it to be fulfilled, and thus baffles all calculation.

–Henri Bergson

How is the emergence of *telos* from chaos to be explained? How is the emergence of *telos* from a merely mechanistic world to be explained? To answer, I turn to pragmatist philosopher and semiotician Charles Sanders Peirce (1839-1914), whose work has found renewed interest among biosemioticians.

Peirce was a major figure in my literature graduate studies, long before I ever heard of biosemiotics or even complexity science. I first encountered his work when I was studying poet Wallace Stevens (1879-1955) with Stevens' biographer Joan Richardson. Some of the difficulties of this famously difficult modern writer disappeared when she assigned Peirce's essays, which Stevens admired and used. It's not as though Stevens regurgitates Peirce's pragmatism, but it does animate many of his lines about chance or order, such as these from "The Idea of Order at Key West" (1936):

She was the single artificer of the world
In which she sang. And when she sang, the sea,
Whatever self it had, became the self
That was her song, for she was the maker.

Peirce's pragmatism, I learned from Joan Richardson, provided not only an understanding of poetic forms, but an understanding of form itself. In these seminars, I entered complexity and self-organization through the door of literature, the only one that was open to me at the time.

Peirce's understanding of chance was, like mine, intimately involved with his theory of final cause and purposeful behavior. Neither the existence of chance nor the absence of it, Peirce knew, falsifies the notion of purpose. The ability to act purposefully is reflected in the making use of chance (as defined above: constraining relationships that form emergent systems). If chance is not there to begin with, it can emerge through stochastic resonances.

Peirce was an incredibly original thinker. As Louis Menand, one of my other professors at CUNY, notes in *The Metaphysical Club*, Perice's ideas influenced many influential New England Americans, but his own name is relatively unknown. He is a philosopher's philosopher. He is also guilty of writing prodigiously and writing in a style that is sometimes very hard to follow. His work was not properly edited or widely published in his lifetime. Much of it remains unpublished to this day. Some of his obscurity is owing to the fact Peirce had an infamously bad temper and did a few scandalous things in his lifetime—for instance, he divorced a devoutly religious American wife in order to marry a French "gypsy" of questionable reputation. Such things kept him out of respectable society and impinged upon his professional success. But this also forced him to work in relative isolation frequently, and this misfortune may have helped his originality along.

Peirce got work as a "geometer" and "computer" through family connections. His daily life involved not only taking precise measurements and working out probabilities, but it also involved thinking deeply about the nature of precise measurement, objectivity, and probability. Peirce in his inestimable genius supposed that something he called "absolute" chance, or radical unpredictability, existed prior to all matter, space, and time, and was the source out of which all matter, space, and time emerged. Peirce's conception of absolute chance can be compared to what we now know as quantum states, which are not directly determined by past states. Peirce opposed absolute chance to the then current, classical definition of chance, a term used to indicate ignorance of true causes.

Peirce likened his approach to the kind of logical exercises performed by the pre-Socratics when they asked themselves questions, such as: Did the world begin from nothing or from something? Was the original state homogeneous or heterogeneous? The pre-Socratics assumed that while the existence of sameness (homogeneity: all mixed up randomly, without structure, no formalizable variation) does not require an explanation, the existence of variety (heterogeneity: at least partially ordered form, formalizable structure) does. And therefore, they further reasoned, variety must have come from sameness. The ground for the assumption is this: We are not satisfied to suppose that all variety (something) came from other variety (something) for this leads to an infinite regress. If everything comes from something else, what did that something come from?

But what if the original state were one of sameness? Imagine an original state that is more random than white noise. Finding yourself in the midst of it, you would experience no direction that is distinctive: all would be homogeneous in that all would be utterly disordered, no discernible difference anywhere and thus no variety. This would be chaos, a kind of nothingness. Sameness would be *like* nothing or as close to nothing as we can put into words. So if variety (something) came from sameness (nothing), you could stop your search.

In "A Guess at the Riddle" (1887-88), Peirce describes absolute chance as chaos that came before time and regularity. Chaos is pure potential that has not yet had an effect because it is so irregular.

> *The existence of things exists in their regular behavior. ... Not only substances, but events, too, are constituted by regularities. The flow of time, for example, is itself a regularity. The original chaos, therefore, where there was no regularity, was in effect a state of mere indeterminacy, in which nothing existed or really happened.* (278)

Chaos, by Peirce's definition, is that which is utterly homogeneous (or highly entropic) in its *uniform lack* of regularity and predictability. However, it does not have any rules to govern its behavior or to make it continue to behave in an ideally random (non-repeating) way. Thus, even chaos can produce coincidental regularity, a pattern, or what Pierce called a bit of "primal matter." As soon as primal matter (which the PreSocratics called the *arche*) is put into relationship to chaos or to a different kind of primal matter, there is a sense of "polarity" or difference. If this polarity is self-reinforcing or persists, it has begun to "take habit." This is the effect of *telos*, for the cosmos has begun to spontaneously organize itself. It moves from a state of high entropy or homogeneity to a state in which some pattern, structure, differentiation, and heterogeneity exists. Peirce's theory, or rather my version of it, since doubtless I have taken liberties, is not inconsistent with "inflationary models" that are now being proposed to explain how differentiation occurred in the early universe (see Linde). It is also similar, I hope you noticed, to the ╙ and ╥ chemical reaction example I gave earlier. Peirce's "creation myth" also resonates with Judaic mythology that explains creation by differentiation: the separation of darkness from light, but in his theory, there is no one commanding the separation. There is language, however, after a fashion, in the way interacting parts can form signs: indices or icons, and eventually symbols. While there is no one to *say*, *Fiat lux*, perhaps semiosis is nevertheless behind creation.

Peirce imagined that the determinism that arose out of absolute chance was not exactly predictable, not one-hundred percent. He thought that nature's laws were only highly probable, so probable as to be virtually inevitable. Out of the original chaos emerges *probabilistic necessity*, our physical laws, and this gives us determinism. This is essentially what quantum science researchers have found to be the case. The necessity we experience is not prespecified because it arises out of quantum fuzziness.

Peirce nested probabilistic determinism within indeterminism and asserted that order can spontaneously arise out of disorder. This was a *very* remarkable claim for his time. It had been assumed that only order could beget order. *Peirce recognized that the very unpredictability of "absolute" chance, leads to the most predictable kinds of statistical regularities in time.* Although this claim may at first seem counter-intuitive, the logic becomes clear if one considers the kind of systems that exhibit the greatest degree of statistical regularity. For example, although no one can predict exactly when a person will meet his or her end, insurance companies can predict the average age of death with remarkable accuracy. Life expectancy statistics are predictable *because* the individual events that contribute to the average are uncorrelated. We can arrive at a useful macroscopic description only if there is microscopic uncorrelatedness that can be averaged out. If deaths *were* causally connected, that is, if the second death occurred because the first one did, and so on, it would be much more difficult to predict the outcomes (one would have to try to use nonlinear analyses). It would be a little more like trying to predict the stock market where earlier trades have an effect on later trades. Nonlinear effects prevent prediction based solely on statistical mechanics.

Indeed insofar as we know, quantum states are uncorrelated and therefore lead to statistical regularity. It may be that quantum mechanical systems also have nonlinear properties, but at the moment these properties, if they exist, are unknown, and probabilistic descriptions (*e.g.* the *linear* Schrödinger equation) seem to work perfectly well to describe the long-term behavior of quantum mechanical systems.

If Peirce's intuitions are correct, absolute chance has no past, and, therefore, the present cannot be correlated to past events. This is how he explains the purely statistical regularities that an "original chaos" would eventually produce. Although Peirce's sense of probabilities stems from a long tradition (see Hacking), he had the original idea that the statistical regularities do not indicate an unknown and fundamentally true regularity. Rather, he thought the most fundamental condition was absolute irregularity.

So now that we have a narrative to describe how indeterminacy can lead to the beginning of time and space and deterministic laws, we now need to describe how it might be possible to escape from these laws. How would freedom of action be possible in a deterministic world? As hinted above, we need a situation where feedback exists and a group of interacting parts can cohere into a whole.

In "Design and Chance" (1883-84), Peirce imagines that a million gamblers, with a million silver dollars each, gather to play a game of chance using dice. At first the game would follow statistical prediction. At the end of a million bets

> *About ten will have lost $2,000 or more, no one over $3,000; and half of them after playing day and night for nearly a fortnight at the rate of one bet a second will stand within $300 of where they started.*

But then he supposes that the conditions might be changed slightly,

> *...the dice used by the players become worn down in the course of time.... And ... they are worn down in such a way that every time a man wins, he has a slightly better chance of winning on subsequent trials.* (220)

Saying this, Pierce puts his game of chance within a system of constraints. Due to constraints in a complex system, events become correlated: the chance increases that what happens once (no matter how improbable, statistically speaking) will happen again. And so the behavior of the parts becomes more limited and more regular than they would be if each part were in relative isolation.

> *This will make little difference in the first million bets, but its ultimate effect would be to separate the players into two classes those who had gained and*

those who had lost and these classes would separate themselves faster and faster. (220)

In this way Peirce's game of chance moves from deterministic and predictable to unpredictable. This theory of self-organization foreshadows the theories of complexity scientists that would be developed in the 1980s and 90s.

In the seminal essay "Chaos" (1986), nonlinear dynamics theorists, echoing Peirce, proposed "the exercise of will" may be understood as the local structuring of random changes. Advancing on this research, many cognitive scientists, psychologists, and philosophers of mind now say intentionality manifests itself (at least in part) in the peculiar and dynamically stable way a person recognizes and uses patterns found in randomness. This is referred to as *dynamical autonomy* and is aligned with ideas about order as arising out of disorder that are necessarily part of a post-evolutionary, post-quantum mechanical, complexity science understanding of natural processes (Wimsatt, "Ontology").

Peirce's writings and philosophy appeal to me because they offer a way of addressing and resolving the arguments and objections to teleology that both the determinists and the indeterminists throw my way. Unfortunately, if one is grossly misunderstood in her profession and trying earnestly to counter claims that her theories are eccentric, it does not do her much good to choose a grossly misunderstood, eccentric man, such as Peirce, as an authority. I should note that I neither agree with nor understand (nor have read) everything Peirce wrote. On occasions, especially late in his life, he veers off on what seem to me mystical tangents, as for example in "Evolutionary Love" (1893), in which he declares self-organization is nothing less than cosmic agape. Although I did not follow him up that path, in general he has been a great inspiration.

I admire an obscure, misunderstood and imperfect philosopher, who in essence simply restated what the pre-Socratics had already said. I might have invoked instead more reliable theorists who state some of his intuitions in more scientific terms, such as quantum field theorist Lee Smolin. So why use Pierce? Number one, Peirce's theory is also a theory of semiosis, and we will shortly investigate more fully how important this is to self-organization. Two, Peirce's historical significance and influence on art, particularly postmodern art, is great, if indirect. His contemporaries included William James and William's brother Henry James, whose influence on modern literature is inestimable.

Later, as noted, Peirce was admired by Wallace Stevens, a major figure in a famous late Modernist salon, which included Gertrude Stein (Richardson 98-136). I've argued (see "C. S. Peirce's Theory") that Peirce was important to major postmodern writer Thomas Pynchon, and that he is, in fact, the referent to the character Pierce Inverarity (in *variety*) in *The Crying of Lot 49.* The popularly-known Umberto Eco has written on Peirce. Peirce also influenced, as I've already mentioned, Jacques Derrida. In general, Pragmatism, founded by Peirce, is very important to postmodernists.

I've also claimed the postmodernists got Peirce wrong. Actually, in some of Peirce's early work, he *is* contradictory and misleading. Not until he fully developed his theory of final cause did he work out those contradictions. According to T.L. Short, Peirce's early semiotic had three faults, later corrected, which postmodern philosophers embraced: "it makes the object signified to disappear; it makes significance to be arbitrary; and it fails to tell us what signification is" (*Peirce's Theory* 44). These three faults helped, possibly, to contribute to the typical postmodern idea of a "self" as incoherent, unstable, and without purpose. Short claims that Peirce's "mature semeiotic" allows for metaphysical realism, non-arbitrary signification, purpose, and dynamically stable selfhood. My interpretation of Peirce's semeiotic is the subject of the next chapter.

Chapter 5

SEMIOTICS

The more constraints one imposes, the more one frees oneself from the chains that shackle the spirit... and the arbitrariness of the constraint serves only to obtain precision of execution.

–Igor Stravinsky

Semiotics is the *study of signs*. A brief explanation of what is meant here by "sign" is in order. I follow Peirce here. I will also very briefly touch upon the basic ideas behind postmodernism semiotics. Fortunately, the difference between Peirce's "mature semeiotic," as T. L. Short calls it, and postmodern semiotics is clear and easy enough to grasp if you are familiar with semiotics. Unfortunately, those unfamiliar with semiotics may find this chapter a bit trying. However, in a book about teleology, some alternative to radical postmodern constructionism needs to be presented—to those of you that consider yourselves postmodernists, that is—others may be less interested in this chapter.

Peirce thought every sign logically has *three* parts: 1. The representamen (what represents) 2. The interpretant (the response to the representamen) 3. The object (what is represented). Peirce had a liking for awkward terms, and for this I apologize. Describing the three parts of a sign, Peirce shows us how signs work. Every sign involves: 1. Something that is 2. Interpreted to stand in for 3. Something else not present.

Further, he thought it was important to distinguish between three different *types* of signs: icon, index, and symbol. We may think of the types of signs as three different kinds of interpretants to a representamen or, in plain English, as three different kinds of responses to a representation. An icon is similar to its object, the thing it's standing in for. An index is contiguous with its object, physically connected to it in some way. A symbol is conventionally associated with its object.

- An icon represents its object by virtue of a likeness. Birds flocking can represent the same object as fish schooling, in a general sense. Any particular token of a type represents the type by similarity. Any bird represents some aspects of birdness.
- An index represents its object by virtue of being near the object (or part of it) and thus being affected by it. A bird's constrained movements in a flock represent the dynamic entity we call a "flock."
- A symbol is only arbitrarily linked to its object and comes to represent its object through convention or habituation. In the English language, the word "bird" represents birdness.

So, to put it yet another way, the representamen and the object relate either by similarity, contiguity, or arbitrarity. The objects of these representations are always general, never particular; they are types of things, not particular things (see Short, *Peirce's Theory* 117-50). Note above that the examples of objects I give above are all general *types* of things or processes: flocking behavior, a dynamic flock, and birdness. It's important, as far as an understanding of *intentionality* is concerned, to realize that semiotic objects are *types* of things not particular tokens. Only the responses (interpretations) and the representations can be particular. I'll come back to this very important point about the semiotic object later.

SYMBOLS

The symbol may the type of sign that's most familiar to us, but it is the most complicated and difficult to understand. Let's take the symbol "bird." There is no inherent relation between the sound "bird" and birdness; "bird" may not be better than "*Vogel*," the German word for bird. A symbol is an arbitrary sign. The object of a symbol is related to it by habitual association or convention only. "Bird" is just the verbal habit we English speakers have that refers to birdness. The object birdness is a flexible category not illustrated fully by any actual animal.

Even though symbols are arbitrary, they don't just come into being out of thin air. They tend to grow out past relationships, out of icons and indices. The Old English word "bridd" is the original source for our word "bird." Originally referring to *young* birds, "bridd" may have been, according to folk etymology, related to the word for "brood," which is related to an early Germanic term for "to heat," whence also the English "brew." The incubation (heating) of bird eggs by their parents is called "brooding," and the group that hatches is called a "brood." The *Oxford English Dictionary* notes that "bird" is ultimately of uncertain origin, but I would be willing to accept the folk etymological explanation that links brew, brood, and bird by means of *similarity*. To heat eggs is *like* brewing birds. Thus "bird," formerly "bridd," comes to represent a type of animal whose eggs are incubated. Our word "bird" may now be an arbitrary symbol for birdness, but there is probably some reason (if not this one) for the sound "bərd" and not some other. As a symbol now, however, "bird" has lost that direct connection to brewing (if it had one): we don't think "brew" automatically when we think "bird," and so "bird" has become an arbitrary sign for birdness.

A symbol, then, might be defined as a regularity that has become arbitrarily *associated* with something else. Symbols tend to grow out of and to transcend (or negate) previous iconic or indexical associations.[1] A symbol can't be arbitrary unless it negates the icon or index that originally connected it to its object.

Although one might think *symbols* are strictly human signs—because they are so arbitrary and must be developed by convention—we also find habitual associations elsewhere in nature, not just in human language and culture. These too would grow out of and eventually transcend iconic and/or indexical connections. Let me give an example of a symbol in biological processes. When discussing the habits of cells, biologists often refer to *meaningful reactions* as cellular "signals" if they tend to trigger useful processes. The type of signal that may have been selected to trigger this or that process is often arbitrary in the sense that it is not inherently connected to the process it triggers. It may be that different species use different chemical signals to trigger very similar kinds of responses. We may say that different species use different symbols to mean essentially the same thing, much as different groups of humans use different words to mean essentially the same thing. This arbitrariness is fascinating to us, and we are surprised to find symbols used in cells, so surprised, in fact, that we are tempted to say that "signal" is just used as a metaphor in these instances. But as biosemioticians argue, we *can* take them literally. We just have to supply a theory describing how such arbitrary meanings do emerge by transcending iconic and indexical connections.

---

1. Jeffrey Goldstein has noted that emergence must involve negation to allow radical novelty, not merely ordinary change. See "Emergence, Creative Process, and Self-Transcending Constructions."

Supplying such a theory is not that easy. As linguist Ferdinand de Saussure (1857-1913) astutely observed, "speaking of linguistic law in general is like trying to pin down a ghost." I think this is so because the underlying "rules" tend to involve stochastic resonances—as between ╙s and upside down ╥s—as between brewing and incubating—so they cannot be predetermined in any sense and seem, well, really "lucky." A word-symbol for birdness might have come from *any* chance resemblance between the sign and *any* characteristic of aves. Rules might involve formalisms, which parts happen to form patterns because of some chance iconic or indexical connection, and rules might also involve functions, which patterns happened to be useful (self-perpetuating) because of some chance iconic or indexical connection. Perhaps the precursor to the Old English "bridd" came from some other unknown source and had nothing to do with "brewing," but the chance association with brewing made the word seem to fit and it stuck.

Self-organization, in language or in biological processes, entails the existence of such underlying rule-like relationships, which lead, a bit surprisingly, to attractors, archetypes, or common types of patterns. This may be, in my humble opinion, the source of the intuition that there are "deep structures" in language that are universal. But, contra Noam Chomsky, I would say "highly probable" and "emergent" rather than "predetermined" and "universal."[2] This may also be the source of archetypes in animal morphology, so noted by non-Darwinian biologists.

ICONS AND INDICES

When we speak of "signs" in ordinary language we usually mean human *symbols,* but this is merely one kind of sign and a fairly mysterious one at that. Peirce insightfully argued that signs are more various and existed before *Homo symbolicus* appeared on the scene. I have furthered (I hope, and not "plundered") Peirce's argument by noting, in my ╙ and ╥ example, that there are icons and indices in self-organizing reactions: the by-chance like and by-chance near, respectively, that represent a larger pattern or type.

An icon in human culture is any thing that stands in for a type of a thing not present by being similar in appearance to it: the icons on Men's and Women's rooms doors, the printer icon on your desktop. An icon in a biological system

2. See also Terrence Deacon's argument against Chomsky in *Symbolic Species.*

might be a lymphocyte receptor, whose shape is a negative image of the shape of an antigen.

An example of an index in human culture might be a weather vane, according to Peirce, which points in the direction of the wind. Indicative gestures, made in specific contexts, also point to objects. A weather vane or a pointing index finger instantly indicate a trajectory to the viewer. Unlike symbols, indices do not need habituation to make them meaningful. One can interpret them (get information about something else, like the wind or the location of some object) without having any prior experience with them.

Since these are the quintessentially Peircean examples, I mention them, but I actually find these index examples a bit confusing: they seem to contradict my theory that wants all sign-objects to be types, not particular things, like "the wind" as a definite object. I think of indices as contiguous parts of wholes, whose behavior or qualities give an indication of an "unknowable" whole. I insist on this because, as an emergentist, I say that we don't need the concept of the sign in science *unless* there are dynamically stable types and wholes that can *only* be partially known through signs. Instead of a weather vane, I prefer the example of a leaf caught up in a dust devil whose constrained movements indicate the larger pattern in which it is caught. If a person observes (interacts with) a weather vane long enough, he might also get a sense of the larger air current that is making it move in limited, but not completely fixed, ways. This complication of the weather vane example better suits my thinking about how indices work and what they can indicate.

Turning to an example in biological processes, we find an axon firing can be indicative of the complex state(s) of the nerve cell. Other nerve cells can respond to the axon's behavior as a sign of the contiguous nerve cell's emergent state or as a sign of the larger neuronal pattern of which that nerve cell is part. The contiguous nerve cell doesn't have to understand cognitively the sign of its neighbor's activity, the way we understand the movements of a leaf in a dust devil. The nerve cell only has to be affected by its neighbor such that its own responses come to represent the same object, *i.e.* the larger pattern, like the leaf and its neighbors represent the same dust devil. I want to note, however, that when we watch a leaf in a dust devil, our cognitive responses come to reflect the larger pattern of the dust devil, so we may say that cellular and human semiosis differ not so much in kind as in degree of complexity.

All signs, whether icon, index or symbol, represent larger patterns, either by similarity to a type of pattern, by contiguity with a type of pattern, or by previous repeated associations with a type of pattern. Any interpretation of a sign as a sign has the effect of maintaining that larger pattern. Signs are significant to the continuation of the pattern and, as such, *serve the purpose* of maintaining the pattern and themselves. Thus semiosis leads to teleology.

THE SELF-PERPETUATING SEMIOTIC CYCLE

In biology all semiotic processes are clearly purposeful; signs are used to aid in survival and/or reproduction—or, to put it differently, signs are used for the maintenance of the individual's habits, the individual himself, or his species.

A fox smells a rabbit. The object of the sign of the rabbit is not the rabbit itself. The fox's receptors in his nose respond to an odorant molecule as a sign of a type of sustenance; sustenance means self-continuation. Only as sustenance or means of survival does the rabbit have meaning for a fox. The object of any sign is always in some sense the continuation of the interpreter; the object of any sign means self, means self-continuation.

Some readers will object here, noting that *human* purposeful acts are not always so obviously about survival. For example, when we fall in love, we don't do so strictly for sake of reproduction. We don't always go for the strongest, healthiest mate or the one with the most resources. We might go for the one with crooked teeth or funny habits that we find, irrationally, very endearing. Many of the things we do, on purpose, may not be "good" for us. We might gravitate toward mates that remind us of our parents or former mates, even if they weren't particularly good at parenting or mating, respectively. The things we do tend to confirm or repeat the kinds of things we've done and thought before, for good or ill, and in this way tend to preserve the type of people we are. Habits, even bad habits, are a form of self-preservation. Some people can become so locked into narrow habits that they can't effectively respond to things that are different from what they are used to. They may end up in therapy seeking ways to break out of old habits and to develop new habits. We can do this only by reinterpreting old signs to mean new things and to help develop and maintain a new self. To learn to love anew is a kind of adaptation.

Let me turn to a children's tale for help with this intellectual knot that ties together semiosis and self-preservation. In *The Little Prince,* by Antoine de Saint-Exupéry, the prince meets a fox, who wants to be befriended or "tamed." To be tamed, as the fox explains, is to add meaning to one's life that is not merely meaning for the sake of survival or reproduction, but is meaning that is built upon associations and familiarity—habits of interaction—to which one becomes more or less addicted and which begin to *inform* one's identity. To be tamed is to learn to love. In Katherine Woods' 1943 translation, when the fox first meets the prince, he says,

> *To me, you are still nothing more than a little boy who is just like a hundred thousand other little boys. And I have no need of you. And you, on your part, have no need of me. To you, I am nothing more than a fox like a hundred thousand other foxes. But if you tame me, then we shall need each other. To me, you will be unique in all the world. To you, I shall be unique in all the world...*
>
> *..."My life is very monotonous," the fox [continued]. "I hunt chickens; men hunt me. All the chickens are just alike, and all the men are just alike. And, in consequence, I am a little bored. But if you tame me, it will be as if the sun came to shine on my life. I shall know the sound of a step that will be different from all the others. Other steps send me hurrying back underneath the ground. Yours will call me, like music, out of my burrow. And then look: you see the grain-fields down yonder? I do not eat bread. Wheat is of no use to me. The wheat fields have nothing to say to me. And that is sad. But you have hair that is the color of gold. Think how wonderful that will be when you have tamed me! The grain, which is also golden, will bring me back the thought of you. And I shall love to listen to the wind in the wheat..."* (58-60)

The wise fox knows that new meanings can emerge in an animal's life through associations, and these new meanings become part of one's self, as much as the meanings that have been inherited through evolutionary processes. The interpretation of wheat as an iconic sign of the fox's general idea of the prince (not the prince himself) would be purposeful insofar as it would reconfirm the fox's developing feelings for the prince. Signs are meaningful insofar as they continue a way of thinking or being. Seemingly purposeless, human actions and values, like falling in love, have evolved as much by similar self-preserving mechanisms as more obviously purposeful actions have.

So, to repeat, the object of any sign is, in some sense, the selfhood of the interpreter, the continuation of a type of being. We can only recognize what we already know, what is already "us." But this *does not* condemn us to solipsism. We do learn. The way of learning something new is to first mistake it for what you already know. If you take in the thing mistaken as a sign often enough (it tames you) then it can begin to alter who you are. While the fox does not tell the Prince how to *start* the taming process, we realize that they would have to remind each other of things they already know and love.

If the objects of signs were not types of the self but *were* particular non-self things—the odorant stood for the rabbit itself, the wheat stood for the prince himself—then scientific reductionism would be perfectly adequate to describe all phenomena in nature. The odorant molecule would directly, mechanically cause the fox to chase, pretty much as a rolling billiard ball, hitting another, causes it to move. But odorant molecules don't move by efficient causes alone. Yes, odorant molecules physically connect to receptors, but this does not inevitably cause the response of chasing. There must be an *interpretation* first, a physical response that is *associated* with fortuitous effects that have previously resulted in the continuation of the self that smells the odorant.

The semiotic *object*—as general type of self—constrains whatever interpretation can be made of it, *that is*, it is self-perpetuating. The habits of a fox, the habits of his ancestors, define the system that is to be maintained or not, and also define, to some extent, what the fox can "see," "need" and/or "use" or not. He can only see and use potential parts of himself, things that resonate with his identity. According to René Thom, who, as Don Favareau argues, may be the first modern "biosemiotician," prey is perceived by a "phenomenon of resonance, for how can we ever recognize any other thing than ourselves?" (352). All the fox's semiotic interactions in his world will involve responding to things in terms of how they can be used to continue his system, his ways of being. His habitual behaviors and processes and the physical structures that make a fox a fox have emerged in the evolution of foxes. This definition is a restatement, now in semiotic terms, of something I said earlier when I was discussing how the old idea of the teleological "end state" is better understood as the "emergent whole."

Let's go back to that earlier example I gave of purposeful behavior as a cycle, a form of homeostasis, and reconsider—now inserting more terms from

Pierce's semiotics—the quintessential purposeful behavior of chasing prey as a self-organized response (interpretation) to a representation of an object.

Interpretation: fox chases

Representation: rabbit as prey (or as potential part of self)

Object: survival of self, further reproduction

Because the prey represents survival, the fox chases it. The meaning to a fox of a rabbit as "prey" has emerged in the predator's evolutionary history and development. The object in this case is physical organization of the fox that the prey, when caught, is helping to continue. The prey as such (and not as, say, as "pet bunny" as the animal may be to a child) is both created by and helps create the cycle in which it plays a meaningful part.

And here is the illustration translated into an abstract example:

Interpretation: ╚ binds with ╩

Representation: ╩ as ╚ (or as potential part of self)

Object: more ╚s are produced

The clump of ╚s that form a system is only proto-teleological and proto-semiotic (because it can't alter the ╚ response in some way, if necessary or better), but it is precisely these kinds of processes that, when combined with and nested within other like processes in organisms, can result through adaptation in fully semiotic, teleological behavior.

Peirce's triadic semiosis allows us to envision all teleological behavior—which is to say behavior that is *significant* to the continuation or improvement of the entity in question—*in cyclical terms*. The representamen is the rabbit or odor of the rabbit; the interpretant is the response of chasing, and the object is the continuation of the predator's ever-changing but dynamically stable physical organization. The three parts of the sign create meaning by feedback and self-organization. Semiotics and teleology are indivisible. You cannot have one without the other.

I also want to stress here again that all objects of signs are *emergent* objects that can only be known partially through signs. A "fox" is not a particular thing that is so easily defined. While any individual has a skin that partially separates him from his world (and acts as a filter), he is also always changing in that world. And the species fox, like any species, is a type of animal whose particular definition cannot be known precisely. Foxness is an emergent category of animal that is ever changing and maintains a dynamic stability through time. This makes all signification intentional, *i.e. about* emergent objects, about objects not fully present. All objects of signs are emergent wholes, not particular things. We only know the fox by observing its actions, which indicate its foxness.

Postmodernists have praised the dynamic and nonlinear nature of Peirce's *triadic* semiosis as an improvement over Saussure's *dyadic* focus on the signified-signifier, which are roughly the representamen-interpretant in Peirce's system. But what they either missed or declined to adopt is Peirce's teleology and, less obviously, his realism, which is a consequence of his including the (dynamic) object in the semiotic process.[3] According to Peirce, reality is gradually (and imperfectly) revealed in the semiotic process because it includes the object. Although the object may not be a static material thing—in the case above the dynamic identity and fluid structural organization of a particular fox—it is fully part of the semiotic process and the object's effect can be partly known. We see the fox's actions as purposeful and self-sustaining, and thus we "know" the fox, the object being maintained. *The inclusion of the object in Peirce's semiotics explains how there is partial access to complex reality through signs.* While interpretation is fallible, it is constrained by the object. Jim Crutchfield ("Calculi") independently comes to a similar conclusion when he argues that because a part of a system is interacting with other parts of the same system, overtime its behavior can become a *relatively objective* model of the holistic system in which is functions.

In Saussure's semiotic, on which postmodernism is fundamentally based, the conception of meaning derives from the systematic relations between and among signs and *not by referring to real objects*. In this view, wherein signs refer only to other signs—rather than being grounded in purposeful behavior by referring to an object—language *arbitrarily* constructs our notions of reality; hence *postmodern constructionism*. By including the constraining object (the

3. I have learned from Peircean thinker Clark Goble that there is an effort underway by some to (re)define Derrida into a realist. While this may or may not be true of the man himself, whatever works to improve his philosophy seems worth doing. Goble is similarly pragmatic about the project.

purposeful emergent whole, to be maintained *or* developed[4]) in the semiotic process, Peirce connects the semiotic web at much needed anchor points to the real world, the real world *from which the self learns* and is therefore able to inhabit and inform. In this view, the postmodern "problem" of the "arbitrary signifier" is no longer a problem. I would argue that if there are not real objects, there can be no signs, and therefore postmodern semiotics is incoherent from the start. No wonder it was so difficult!

I hope I have not lost too many of my readers. Thanks for sticking it out through this section on Peirce, semiotics, and postmodernism, any one subject of which is enough to make the reader's book slowly slip from his relaxed grasp.

I also want to take the opportunity to give my remaining readers a bit of advice about secondary sources. If you want to know what Peirce said, you have to read Peirce. In this work, I give you my interpretations of Peirce's work. Sometimes, when one is thinking, one needs to change information to make it work. Nature certainly doesn't care about correctness when she is reinterpreting some structure in a new way. Whatever works, works. I believe it is through the engine of mistaken impressions—so long as they are sufficiently constrained!—that progress is often made. I'll discuss more of this anon, much more, since purposeful creativity by means of chance interpretation is the subject of this work.

4. *i.e.* directionality *or* originality.

PART II

HISTORY: SELECTED FIGURES

Chapter 6

DIRECTIONALITY AND ORIGINALITY

It is not the strongest of the species that survives, nor the most intelligent that survives. It is the one that is the most adaptable to change.

–Charles Darwin

Today's misinformed prejudice against "teleological art" is unfortunate because teleology provides extremely subtle tools for artists to explore how one can create something *truly new and yet meaningful.* In particular, teleology has shown that the concept of purpose entails two distinct kinds of behavior that are important to understanding art as such. The first is associated with what I call *directionality*, or the mechanisms for maintaining order: *stability, habituation, automation.* Directionality results in types, species, genres, and habits. The second is associated with what I call *originality*, or the discovery of wholly new functions: *change, creativity, freedom*. Originality begins the process of creating new types, species, genres and habits. Both directionality and originality emerge from selection processes. With *directional* selection, slight differences are inconsequential, and sameness prevails. With *original* selection, slight differences result in significant functional improvement and/ or systematic reorganization. Let me further elaborate on these two concepts and place them historically.

The ideas behind the terms "directionality" and "originality" occur again and again in teleology-focused biology. Certain biologists have been interested primarily in directionality, others in originality. Teleologists often divide along these lines.

Some biologists have noted that there seems to be an internal cause that guides both individual development and species adaptation. What's known as the *directional*-internalist theory of change suggests that intrinsic factors drive evolution in limited directions (see Gould, "Eternal Metaphors"). Nature seems to prefer certain themes and produces various versions of archetypes. Biologists working in directionality are interested in trends, *but not necessarily progress.* They seek to understand the idea of "natural" types and are more interested in understanding why certain forms are common in nature, even if they are not necessarily more functional than less common forms.

Other biologists interested in *originality* have been more concerned with the way nature creates new entities/systems that seem to anticipate unpredictable *future* needs. These biologists are interested in useful innovations and diversification. They want to understand how an organism might break out of the "natural" type, finding a less common form that is more functional (see Reid).

The distinction between directionality and originality is perhaps partially derived from Immanuel Kant's conceptions of the aesthetic and the teleological judgments, which informed the arguments between two types of 19th century teleologists called transcendental morphologists (concerned with form) and Kantian teleomechanists (concerned with function). Formalists are interested in the proportional relationships of the parts to the whole, with the purely aesthetic design and organization without regard to function or meaning. Functionalists are interested in how the parts function vis-à-vis a whole. They want to know how something works and what it means for survival. Today the division continues as the formal concerns of "neutral evolutionists" or "structuralists," such as Walter Fontana, Brian Goodwin, Stuart Kauffman and Kalevi Kull are opposed to the functional concerns of Neo-Darwinists, represented at their extremes by Richard Dawkins.

Originality, being associated with function, is clearly linked to *final* cause, *i.e. purpose* as cause. Directionality may be more clearly linked to *formal* cause and to self-maintenance than to final cause *per se*. Yet formal causality is necessary for the creation of form that can later take on teleological meaning (or not), that is, become useful or not. Moreover, formal cause/directionality has been associated with final cause and purposefulness for other good reasons, since with directionality, the parts *serve the purpose* of self-creation and self-maintenance of the whole, and exist, in the whole, for that *purpose*.

Often it may be unclear whether a particular teleologist has a theoretical preference for directionality or for originality. In some cases, theorists assuming this or that label may argue against oppositely-labeled theorists when in fact they have much in common. In other cases, theorists wearing the same label may not be talking about the same thing at all. To give a good example of the extent to which these issues are confused, the 19th century German teleomechanists studied directional or formal cause phenomena but were self-proclaimed "finalists," united against Darwin, whose theory may be considered a naturalization of final cause.

This divisiveness and confusion has led evolutionary theorists to waste a lot of time talking past each other. Few teleologists have argued with passion that telic phenomena must involve *both* aspects. But originality can only occur against a background of directionality. And original events (*e.g.* stochastic resonances) can lead to directionality. Understanding this removes any objections critics may have about teleological art being either too conventional *or* too strange, for by this definition, purpose involves radical *and* yet meaningful creativity. If I contribute anything useful at all to this already very long and complex discussion about purpose in nature, it may be my emphasis on these dual aspects of teleology and the need to incorporate them both into a theory of purpose.

Among the few who have specifically noted this dual aspect is Jean Baptiste Lamarck (1744-1829). According to Stephen Jay Gould, Lamarck has been misremembered as being primarily responsible for promoting the "acquired characteristics" theory of evolution; he was merely including what was more or less an accepted notion of his day. Lamarck's actual theory included *two* mechanisms of evolution: one involved an inherent tendency toward complexification, and the other involved accidents of history, as organisms dealt with their environments (*Structure* 170). But Lamarck was perhaps a little too vague about what these mechanisms actually are.

English biologist Richard Owen (1804-1892), who is credited with identifying the first "Dinosauria" fossils as extinct species of reptiles, noted teleology's dual aspect with more specificity. He was a pre-Darwinian evolutionary theorist, who began his career believing, not in descent with modification as Darwin would, but in some sort of variation on divine themes. He conceived of two principles operating in evolutionary processes: one, which he considered "directional," brings about stability and similarity or "a vegetative repetition of structure"; the other, which he called simply "teleological," brings about change and diversity by shaping an organism or plant according to function. He claimed the former (directionality) is illustrated in the conception of a *groundplan* or *archetype* and in the mathematical symmetry of some organisms and even crystals. The latter (which I call originality) is illustrated in the conceptions of adaptation and *progressive* complexification. Owen argued that directionality results from a *polarizing force,* which produces similarity of forms across species, while originality results from an adaptive *special generalizing force,* which produces the diversity of organic forms (Russell 102-12).

Similarly, the lesser-known German anatomist C. B. Reichert (1811-1883) made a distinction between "class" (aka directional) and "functional" (aka original) modifications in the structures of biological organisms. While Reichert could not explain the cause of the former, he argued that the latter occurred in response to the environment (Russell 145-7).

In contemporary theory, the duality of teleological phenomena has been noted in, for example, Stanley Salthe's distinctions between development and evolution, Erich Jantsch's confirmation and novelty, and in Jim Crutchfield's notions of intrinsic evolution and extrinsic evolution.

DIRECTIONALITY

Nineteenth century transcendental morphologists were mostly concerned to account for the differences and relationships between species. They were interested in directionality, in the neutral interrelations between parts and wholes, in how, for example, the change in size in one organ might be correlated with a change in size in another organ. They believed that the direction of evolution was partly predetermined, in the sense of limited, by *intrinsic* factors rather than by selection pressures to be found in the environment. Another term for this is *orthogenesis*. Orthogenesis has been confused with the idea of divinely "directed" evolution, but these theorists did not think of the direction as having a predetermined goal. They thought that there were inherent drives that made evolution proceed in more or less in a straight line—having tendencies—instead of branching off randomly. "Ortho" means straight. Directionality may be a tendency, but that tendency is not toward perfection in any sense. It often just means more and more of the same thing, like bigger and bigger antlers, for instance (Reid 267-287, 392).

Although the transcendental morphologists' claims were rejected after Darwin, nonlinear dynamics theorists now argue that there *are* relatively few archetypical kinds of patterns that result from self-organizing feedback processes. This leads to the appearance in nature of what these teleologists called "variations on a theme." For example, similar laws guiding biological development produce "gills" on mammal embryos as well as gills on fish. It may *not* be that we all share a common aquatic ancestor from which derives this similarity: it may be that the self-organizing processes in early development tend to make these forms, regardless of whether they were or could be useful in water or not. St. George Jackson Mivart (1827-1910) pointed out that although marsupials of Australia have evolved separately from European placental mammals under different environmental conditions, there are marsupials that are strikingly similar to wolves, cats, mice, and squirrels (see Goodwin 23). Are the similarities between a marsupial mouse-like animal and a mouse merely analogous? Or are the similarities due to similar self-organizing processes and laws of pattern formation? Variations on a theme?

Today structural evolutionary theorists claim that if there were a film version of Earth's evolution that could be rewound to slightly different initial conditions and run again, many of the forms we know today would reappear (Fontana). By the same argument, we may suppose that other planets in the universe might be home to creatures much like those on our planet. This goes against arguments by evolutionary theorists such as Stephen Jay Gould who contend that if the "tape" of evolution could be rewound and started again, with just slightly different initial conditions, vastly different forms would result (*Wonderful Life*).

ORIGINALITY

While the transcendental morphologists were primarily interested in directionality and similarities between species (how individuals conform to an ideal), other teleologists were interested in the differences between species (why individuals depart from an ideal). These teleologists considered the role of function in determining the morphological changes in animals. They were interested in the *original* aspect of *telos*, which involves how an organized whole or neutral pattern (produced by directionality) is *used* or *misused* by another system to its advantage. Work in this field was, of course, taken over by Darwinian evolutionary theorists.

The original aspect is apparent when evolution takes a new, unforeseen and unforeseeable direction. For example, the origin of complex eukaryote cells (cells that contain nuclei and organelles) may be viewed as a series of lucky coincidences that turned out to be beneficial for all involved parties. Eukaryote is a term for all organisms included in the kingdoms Protista, Plantae, Fungae, and Animalia. You and I are eukaryotes, and we wouldn't be here if it weren't for some common mistakes (in the sense of new behaviors that were neither inherently determined nor instinctual) that occurred long, long ago. Lynn Margulis, after K. Mereschkowsky (1855-1921), argues that the little organs, or the "organelles," within eukaryote cells, are descended from free-living organisms, known as prokaryotes, that were consumed by a larger prokaryote.

The moment of first contact between a proto-organelle prokaryote and its host (also some sort of prokaryote) must have occurred as a result of a coincidental meeting. The story begins with a prokaryote whose round shape had become invaginated. This was probably the result of some law of form, having to do with the way surface tension maintains an enclosure. Beyond a certain size threshold, the outer layer of the cell caves in, creating a mouth-like area. Since prokaryotes "digest" food by releasing enzymes into the immediate environment, if food got caught in the newly created mouth-like area, enzymes would be used more efficiently because they would not float away so easily. This kind of innovation might occur independently in a number of unrelated organisms: natural selection's gradual "shaping" force is not needed to shape such a mouth. The spontaneous invagination creates a radically new functional structure, a "mouth," quite by accident. The next major development in the plot of this story would come when another prokaryote got caught in this prokaryote's "mouth."

When the two distinct organisms met, one engulfed the other but did not digest it. The second, now inside the first, survived easily because it received a constant supply of food. The first also benefited from the chemical by-products of the second, which improved the internal processes of the first. As Wendy Wheeler has noted, this "failed act of eating" led to an innovation much more valuable than the energy that the engulfed prokaryote would have provided. The engulfed prokaryote had originally meant "food" to its engulfer, but that meaning changed. After time, they adapted to each other, as any two organisms will that are coupled. The one inside became simpler because its new environment was more predictable, but it also came to depend more on the one that housed it. After time the newly created composite system stabilized. Eventually, their descendant's cell divisions became synchronized and they became one functional unit, a eukaryote cell.

The stochastic resonances described in Chapter Two result when random patterns are interpreted as a meaningful, either confirming a directionality or instigating a new directionality. Similarly here, functional reinterpretations and creative misinterpretations of one system's structure by another system resulted in new meaning. In this case, the accidental functionality discovered by the engulfed prokaryote and its engulfer becomes a cause for the composite system's continued existence.

Dorion Sagan offers another example of stochastic resonance. Here life is a "poor artist" who cannot afford to buy new materials and must work with what chance throws her way.

> *Human teeth, for example, are converted toxic waste dumps: evolutionarily, my teeth derive from the need of marine cells to dump calcium waste outside their cell membranes. Calcium is a mineral that will wreak havoc with normal cellular metabolism. Trucking this hazardous waste across cell lines in ancestral colonies of marine cells may well be the basis of all present-day shell- and bone-making.* (46)

In such cases, a *fortuitous benefit* of a property created "accidentally" by physical constraints becomes a *genuine adaptation,* as Peter Godfrey-Smith might say, a property whose *continued* existence is not accidental but explained by its being useful to the organism. The result, in retrospect, is designed (yes, I use "design" because this process is how designing gets done, see below, not from a predetermined plan). In this way, products of nature are works of art.

DIRECTIONALITY AND ORIGINALITY AS VALUES IN ART

Just as truly teleological phenomena must involve both directionality and originality, so must art be made from self-organizing forms harnessed to create ever more complex nested levels of meaning and interaction. A teleological entity maintains its self by making selections from the environment, turning its surroundings into itself, as it were. This is directionality. But it is also partly changed by environment and becomes "like" it, the more it is able to respond to it in diverse ways. Organisms learn. This is originality. Novels, poems and paintings, likewise, become more complex as they are created, as their own patterns become their own constraints determining further selection, but as selection incorporates stochastic resonances they develop further patterns and new constraints. More complexity means more interpretive flexibility for the growing artwork itself and later more points of entry for the audience. Art and audiences learn.

Literary "new critics" C. K. Ogden and I. A. Richards are among those who have lately observed that poem is self-organized. Cleanth Brooks specifically refers to "the structure of the poem as an organism" (218). But the idea is an old one, one that is sometimes awkwardly pitted against the idea that poems are *art*ificial.

I argue that nature and art are not opposed insofar as both are self-organizing. The random details of a long poem, say, must come to form a pattern through a selection process that is constrained by the artist's habits or themes—patterns involving what's near what, what's like what. Themes in art are similar to biological forms, which tend to be stable within limits. Beyond a certain threshold, a new system of organization may emerge. As, for example in poetry, when the Christ pattern emerged in Blok's *The Twelve* and changed the soldiers into martyrs.

A more detailed example of a simple likeness (originality) evolving into part of a theme (directionality) is offered by Brooks (30-32). The process begins with an accidental likeness: a homonym is selected, then an accidental functionality is discovered for the selected form, and then it is fitted within a larger overall theme. In Shakespeare's *Macbeth* the (anti) hero stabs king Duncan then pretends to have discovered the murder. He says,

> *Here lay Duncan,*
> *His silver skin lac'd with his golden blood;*
> *And his gash'd stabs look'd like a breach in nature*
> *For ruin's wasteful entrance: there, the murderers,*
> *Steep'd in the colours of their trade, their dagger*
> *Unmannerly breech'd with gore…* (2.3.109-114. qtd. in Brooks)

Brooks points out that the metaphor of daggers "breech'd with gore," which is to say, daggers wearing blood-red pants, has seemed to some a poor choice, an inappropriate image. Possibly, while Shakespeare was searching for an image for the way the daggers might have looked, the earlier "breach" suggested "breech'd" in that way the mind has of relating things coincidentally but sometimes irrelevantly like each other. Without worrying about how Shakespeare might have chanced upon this phrasing, Brooks argues that the breech'd image is perfectly appropriate to the garment symbolism that coheres at a larger level of the play. Following an earlier critic Caroline Spurgeon, Brooks observes the play has numerous references to dressing and dissembling. The daggers were, after all, dipped in blood artificially to implicate their owners instead of Macbeth, the real murderer. The ultimate connotation of the garment theme is that Macbeth is guilty of trying to put on king's robes that he is not fit to wear. This is how we are meant to regard this tragic hero, and the various garment references and imagery that recur again and again throughout the play help reinforce and call our attention to this interpretation. While "breech'd" might have been one of those accidental associations initially, if we believe Brooks, it resonated with dissembling themes that were then forming in the poet's mind. I agree with those who find the breech'd metaphor a somewhat awkward one, not Shakespeare at his best. But, it is precisely the awkwardness of this metaphor that compels us to search for its stochastic origins as explanation and justification. Aligning breeched with breached is definitely the work of a punning mind: it isn't convention. If "breeched" were a common synonym for "sheathed" then it would have made, as a dead metaphor, a less jarring image.

I think we can guess that its accidental appropriateness to a larger theme is what made Shakespeare want to keep it, against his better judgment, perhaps. Brooks' analysis illustrates well the joint workings of originality (which often doesn't work on all levels) and directionality. The genius of Shakespeare is his stochastic resonances usually work on so many multiple levels simultaneously as to stagger the mind.

If a work is merely directional, as genre fiction is, it may be too predictable and conventional to be considered "art." Individual generic works are products of "Fancy," as Coleridge would say, not "Imagination." Detective fiction, confessionals, pastorals, romance, or thrillers are defined as such by the way they conform to certain conventions. If you know the genre, you can predict, more or less, how things are going to turn out. However, the genre itself as a species is an emergent form, created by means of stochastic resonances. That is, genres or conventions emerge out of the interactions of various writers, all of whom follow their own local rules that they believe (or misbelieve, consciously or not) define the genre. There is no single standardized set of rules to guide them. Thus genres are, from an evolutionary perspective, teleological. However if we are looking at a single instance of a genre, it may appear formalistic, without fresh originality, which we need for art.

If an individual work is completely original, like a dream, (mis)interpreting and refunctioning forms in an *inconsistent* manner, avoiding any kind of holistic coherence, it is has little or no directionality (no convention) and can be unintelligible. James Joyce's *Finnegans Wake* is infamously such a work, and its artfulness can only emerge with the directionality that comes from the interactions of its various interpreters, critics and readers.

Here is a sample quote from *Finnegans Wake:*

> *The answer, to do all the diddies in one dedal, would sound: from pulling himself on his most flavoured canal the huge chesthouse of his elders (the* Popapreta, *and some navico, navvies!) he had flickered up and flinnered down into a drug and drunkery addict, growing megalomane of a loose past. This explains the litany of septuncial lettertrumpets honorific, highpitched, erudite, neoclassical, which he so loved as patricianly to manuscribe after his name. It would have diverted, if ever seen, the shuddersome spectacle of this semidemented zany amid the inspissated grime of his glaucous den making believe to read his usylessly unreadable Blue Book of Eccles,* édition de ténèbres, *(even yet sighs the Most Different, Dr. Poindejenk, authorised bowdler and censor, it can't be repeated!) turning over three sheets at a wind, telling himself delightedly, no espellor mor so, that every splurge on the vellum he blundered over was an aisling vision more gorgeous than the one before t.i.t.s., a roseschelle cottage by the sea for nothing for ever, a ladies tryon hosiery raffle at liberty, a sewerful of guineagold wine with brancomongepadenopie and sickcylinder oysters worth a billion a bite, an entire operahouse.* (179.17-35)

Comparing Joyce to another "difficult" writer, Nabokov, novelist Martin Amis corroborates the observation that Joyce can sometimes deny his readers his principles of relation, his habitual themes; whereas Nabokov's principles of organization are right on the page:

> *Nabokov wants to embrace his readers ... he is the dream host, always giving us on our visits his best chair and his best wine. What would Joyce do?*

Let's think, he would call out vaguely from the kitchen, asking you to wait a couple of hours for the final fermentation of a home-brewed punch made out of grenadine, conger eels and sheep dip. ("On Nabokov")

An artist wants to estrange language but go too far and it ceases to signify. Art must lie in an ever-shifting zone between convention and madness. It must be original, reinterpreting its own directionality by means of some stochastic resonance. It must also be directional, having some kind of conventional or habitual formal order. That order should preferably exist in the text itself, without the help of interpreters bringing the principle of organization in from the outside.

Another way of describing the difference between directionality and originality is to express it as the difference between "non-mentalism" and "mentalism" which is sometimes noted in teleology (See Bedau). *Non-mentalists* are concerned with directionality and internal principles. *Mentalists* are concerned with originality and external principles. A simple analogy—which I have borrowed and adapted from the Santa Fe Institute's quantum physicist-cum-complexity scientist, Murray Gell-Mann—may help to illustrate the distinction. The sequence 3, 6, 9, 12, 15, 18, 21, 24, 27 is governed by a rule. That rule can be discovered by examining the sequence itself: the rule is, each number is increased by 3. This roughly illustrates an inherent principle of organization or rule (which tells one how the sequence might be continued). I also note that the rule is rational and therefore seems like something produced by intelligence, but it is a kind of self-organizing rationality, like a fractal pattern that is repeated at larger and larger scales, and so one need not posit an external intelligence to consciously orchestrate this design, only a selection process, which occurs wherever there is difference and repetition. This sequence may be likened to the patterned beauty of the purely formal effects of literature, which are intuitively felt by the reader. One can appreciate such patterns for their own sake since they do not signify something outside the text.

The sequence 14, 23, 28, 33, 42, 51, 59, 68, 77, 86 is also governed by a rule, but its rule cannot be discovered by examining the sequence itself: the numbers represent stops on the Lexington Avenue subway line in New York City. The rule for this sequence is imposed on the sequence from without. Many difficult works make demands on audiences in this way, containing patterns that signify objects not included in the text, or alluding to obscure texts, like medieval pamphlets, as Joyce does for example, that even the well-educated reader cannot be expected to know off-hand. This kind of pattern requires an external intelligence to reveal the meaning since the pattern signifies something outside of itself. The reader can only make a wild guess at the meaning.

The patterns of non-mentalism are more obvious and do not require interpretation. Here are some more examples of patterns that some feel exemplify something *internally* intentional or directional. I got these examples from an email that was going around attempting to illustrate divine beauty in mathematics. About this one the emailer attached the note, "Brilliant, isn't it?"

$$1 \times 8 + 1 = 9$$

$$12 \times 8 + 2 = 98$$

$$123 \times 8 + 3 = 987$$

$$1234 \times 8 + 4 = 9876$$

$$12345 \times 8 + 5 = 98765$$

$$123456 \times 8 + 6 = 987654$$

$$1234567 \times 8 + 7 = 9876543$$

$$12345678 \times 8 + 8 = 98765432$$

$$123456789 \times 8 + 9 = 987654321$$

This one was noted for its "miraculous" pattern:

$$9 \times 9 + 7 = 88$$

$$98 \times 9 + 6 = 888$$

$$987 \times 9 + 5 = 8888$$

$$9876 \times 9 + 4 = 88888$$

$$98765 \times 9 + 3 = 888888$$

$$987654 \times 9 + 2 = 8888888$$

$$9876543 \times 9 + 1 = 88888888$$

$$98765432 \times 9 + 0 = 888888888$$

Many mathematical patterns have been noted historically and related to mysticism. Fibonacci sequences and amicable numbers fall into this category. Such patterns may be called aesthetic rather than teleological, to use Kant's distinctions. They may also be likened to art for art's sake. But why should any pattern be considered miraculous or divine? Why is it sometimes assumed that patterns cannot form on their own without a conscious agent? It could be that this is the initial source of agency, patterns that form on their own, which are as inherent and unavoidable as these are in mathematics. Formal and neutral patterns first emerge through self-organization, then they can be interpreted, found useful, as signs of something else.

- There is a relationship between non-mentalistic teleology and *telos* as an intrinsic cause (*e.g.,* as a guiding principle or inherent law).
- There is a relationship between mentalistic teleology and the association of *telos* with an extrinsic cause (*e.g.,* a divine creator or human author).
- Non-mentalism can be associated with instinctual or robotic behavior or automata.
- Mentalism can be associated with supernatural abilities.

I think artists want both patterns that are intuitively felt, whose beauty can be appreciated for their formal qualities alone, and patterns that signify meanings beyond themselves. It is in this latter kind of pattern that literature takes on what so many have referred to as a kind of ghostly quality, with patterns seemingly inhabited by an intelligence from another space and time, outside the text. In both cases, non-mental and mental patterns, an author's touch is felt, but the lawful regularity of the former makes it less provocative, I think, than the latter's more mysterious, less objective pattern, whose correspondence may or may not be correctly guessed.

A teleological writer like Nabokov might include many "non-mental" sequences whose principle is self-contained. For instance, the number "342" recurs several times in *Lolita.* It's the Haze home address. It's a hotel room number. Together Lolita and Humbert visit 342 motels/hotels. But I do not think 342 gestures beyond itself at some other meaning. Humbert says it's just indicative of "McFate" that delights in formal patterns. McFate is an aesthete, uninterested in moral problems, and though it may be keeping watch upon Humbert, it makes no judgment and does not offer a meaning of the pattern.

An example of a more evocative self-similarity might be the Lolita-like females in Nabokov's oeuvre: Margot (*Laughter in the Dark,* 1932) who at sixteen is involved with a much older man, Zina (*The Gift,* 1935-37) whose step-father has married her mother in order to be near the daughter, and the Enchanter (in the 1939 short story by that name), whose step-father molests her. These avatars and the later Lolita are also connected to external references that Nabokov brings into the text—Edgar Allen Poe's "Annabel Lee" and a real-life abducted girl named Sally, on whom Humbert mournfully reflects, "Had I done to Dolly, perhaps, what Frank Lasalle, a fifty-year-old mechanic, had done to eleven-year-old Sally Horner in 1948?" (*Lolita* 289). These likenesses of Lolita help guide our interpretations of her pattern and see Humbert's actions as abuse, and thus the pattern becomes more mentalistic than non-mentalistic. That is, for example, Humbert's comparison of himself to Lasalle is judgmental and self-condemning.

But throughout most of Humbert's tale, he avoids self-reproach and tries very hard to cultivate the attitude of the aesthete: he tries not to read into patterns, and he prefers to see them as non-mental, as beautiful formal repetitions. In this way, he reminds me of Thomas Hardy's protagonist in *The Pursuit of the Well-Beloved* (1892), who marries successive generations of girls/women who remind him of his first love Avis. Reading Hardy, I kept wondering, What is he saying with these repetitions? What is the point, other than to dazzle me with formal delights? Non-mental patterns always make a reader wonder, Is there something more?

I think there is. Ultimately, *The Well-Beloved* becomes a critique of Petrarchan conventions, which remove the beloved from real-life context, aestheticize and dehumanize her. Petrarch's Laura is, one might say, the larger system in which both Nabokov and Hardy became part. The title of Nabokov's last and unfinished work, *The Original of Laura* (2009), would seem to support this.

Martin Amis goes a bit further with Nabokov than I do, pointing not just to evidence in Nabokov's work but to history, and makes Humbert—and all Nabokov's abusive males—into icons of Soviet tyrants who effectively spoilt paradise for the young would-be Russian poet, who was forced to leave his home prior to WWII and had to struggle to write in English (*Koba*). I think there is enough in Nabokov about paradise lost to argue that Amis' connection is grounded. And I would go even further to say what makes Amis' connection critically interesting is the fact that we can't be sure he's right, not in the way that we can feel that, yes, the image of Lolita is refracted throughout Nabokov's fiction. There is only the ghostly signification of a tyrant icon, which might be just an accidental likeness, a spurious analogy, which has not been iterated sufficiently enough to become what might be called a conventional Nabokovian symbol in the way that the nymphet is.

In sum, mentalisitic patterns do, in most cases, require an agent to bring some kind of judgment upon the pattern, to say how it functions in a context. Non-mental patterns can form on their own, without conscious agency. This does not stop readers and interpreters from suspecting that there is an agent responsible for non-mental patterns.

Having made distinctions between these two types of patterns, now allow me to blur those distinctions. Non-mentalism/directionality involves the *accumulation* of the effects of stochastic resonances, which we can also call analogies, as for example the way an upside-down ╦ is like an ╚ in certain contexts. Mentalism/originality involves a *singular use* of a stochastic resonance. We may think of mentalism as requiring an agent, with a brain, who is capable of linking together ideas that are physically separate in space and time. Nature, which does not have a brain, must wait upon the coincidence of analogous things coming near enough to each other, like ligands and receptors, to have an effect. Nevertheless, we may say that any selection process that makes use of analogous structures is mind-like. This is probably what C. S. Peirce meant when he infamously declared that intelligence is everywhere in the cosmos.

Chapter 7

ANALOGY AND AFFINITY

I met [my wife] in somewhat odd circumstances, the development of which resembled a clumsy conspiracy, with nonsensical details and a main plotter who not only knew nothing of its real object but insisted on making inept moves that seemed to preclude the slightest possibility of success. Yet out of those very mistakes he unwittingly wove a web, in which a set of reciprocal blunders on my part caused me to get involved and fulfill the destiny that was the only aim of the plot.

–Vladimir Nabokov, from *Look at the Harlequins*

If there is only an analogy between the big and little dippers, there is an affinity between spiral galaxies. An analogy can be defined as a specious comparison between two physically unrelated things. An affinity can be defined as a resemblance in structural organization between, for example, cloud formations, biological groups or languages, implying either a common origin, similar histories of functioning, or reliance on similar developmental principles. To confuse an analogy with an affinity is not sound reasoning or good science, according to most philosophers since the early modern period. Whenever what we believe is a mere analogy begins to appear to have some actual affinity, it seems absurd or uncanny. What if, for example as one of Nabokov's characters once worried, it turned out the "transcendentalism" had something to do with "dentistry,"? ("Ultima")[1] As Freud remarked, the uncanny often refers to the feeling that previously surmounted animistic views of the world—views that mere thoughts have the magical power of linking physically unrelated things—begin to seem feasible again (249).

A primary criticism of teleology is that its method of reasoning is analogical, comparing the way nature creates plants and animals to the way

1. The Latin *scendent*, present participle of *scendere*, (to climb) is unrelated to the Latin *dent* (teeth). That both words, "transcendental" and "dental," coincidentally contain the sequence of letters d, e, n, t, a, l does not imply common meaning.

humans create tools and works of art. Theologian William Paley (1743-1805) is famous for such arguments (for God) from design. He pointed out that we do not wonder how, say, a stone has come to be if we find it lying in a field, but were we to find a watch instead, a watch with parts obviously organized to serve the purpose of keeping track of time, we would be bound to infer that it is a product of intention. By analogy then, a living organism is obviously organized to serve the purpose of its own survival; thus, so argues Paley, we are also bound to infer an intentional creator.

If the discoveries of science—natural selection, self-organization and other kinds of evolutionary mechanisms such as drift or symbiosis—allow us to remove God from Paley's analogy, making nature itself intentional, critics will respond by saying that nature cannot think; nature cannot plan ahead. Even if the ecosystem as a whole does seem self-organized, it still cannot be compared to a person who has a localized center (a brain) that can direct and control actions.

But what if it turned out that we were wrong in thinking that humans have localized centers that direct and control actions? What if it turned out, as I have been arguing, that intentional human behavior is not very different from self-organized behavior that occurs in nature? What if formal selection processes in nature had an affinity to formal selection processes that result in thoughts? What if our conscious deliberations were seen just as our much more sophisticated way of editing ideas already formed by self-organization, speeding up creativity but not creating *per se*? Would the doctrine of teleology be worth looking into again? As Albert Einstein is often noted as saying, If at first the idea is not absurd, then there is no hope for it.

PUSHING THE BOUNDARIES OF THE TERM "ANALOGY"

Analogies that cannot be said to have an affinity are those with merely a coincidental logical similarity, such as the Scorpio constellation and an actual scorpion. The shapes may be similar, but they do not share a common physical origin, nor is the process by which they were created describable by similar mathematical models. Therefore, the resemblance can only be attributed to the subjective intention of the observer. It is the burden of empirical science to make clear distinctions between affinities and analogies.

However, merely analogical resemblances can be significant to the interests of science if they occur in nature time and time again. To take an obvious example, if an animal coincidentally looks like its environment (via a stochastic resonance) and thereby tends to escape predation, its chances of having the opportunity to reproduce organisms like itself are increased.[2] Over time, animal camouflage has developed a physical, ancestral affinity with its environment as genetic information is passed on to the offspring. The affinity develops because a number of would-be predators draw similar analogies between similar animals and their environments.

Directionality (involving repeated coincidences that confirm a direction) is associated with the notion of affinity and is found in *useful predictable* processes, such as thermoregulation and self-reproduction, and the stability of species or true-to-type reproductions. Originality (involving a singular coincidence that instigates a new direction) is associated with the notion of analogy and is found in *useful unpredictable* processes, such as the adaptive evolution of species.

ARISTOTLE ON ANALOGICAL REASONING

Based on the arguments given above, we may say that teleological phenomena can be attributed partly to the analogical reasoning that nature and time perform through the process of lucky selections (neutral or based on differential fitness), which result in either formal or ancestral affinities. From our perspective where we can recognize the ordering tendencies of chance that result in telic phenomena, it is ironic that in *Physics,* Aristotle tried to disassociate chance from his teleology. He argues that coincidence and chance are not to be reckoned among causes because they do not happen *always or usually*. He adds that although a coincidence might lead to effects, any one particular occasional effect would not come to much (Bk. 2 Pt. 5). And it is true that the singular coincidence tends to go unnoticed, and all that is obvious is the directional pattern that has emerged over a period of time.

Aristotle was a non-mentalist. He disapproved of those who tried to involve analogy in teleological explanation. In his non-mental approach, as

2. Camouflage does not actually look *like* its environment. Technically, it is more correct to say a camouflage pattern has a similar degree of regularity/irregularity as found in the environment. See Alexander, "Nabokov, Teleology and Insect Mimicry."

Mark Bedau has noted, Aristotle radically departed from Plato and differed as well from many predecessors (62). Plato could only link final cause with external mind. In *Timaeus*, he insists that final causes "work with intelligence to produce what is good and desirable..." (46.e.57).

Aristotelian teleology, as found in his *Physics*, attempts to explain directionality, or how and why a system maintains itself without any *external* directing agent or principle. Aristotle's physics did not posit a supernatural agent external to the sphere of action who occasionally intervened, creating new regularities that went above and beyond usual law-like events. Rather, he argued that an already existing inherent principle of organization permitted physical processes to run a determinate course. Aristotle's picture of the universe is best described as an immense interrelated organic machine, a bit like James Lovelock's image of Gaia, that is finally determined by the "Idea of ideas," the First Cause, which I, parting Aristotle's company, would call the ordering tendencies of chance. But Aristotle does not note, as Lovelock does, the extent to which chance associations determine direction.

However, importantly, Aristotle does note that chance and *telos* were often confused. In *Physics* he relates an anecdote about a man who went to the market for the purposes of earning money and happened to run into someone who owed him money (Bk. 2 Pt. 4). Because the chance event happened to fulfill his purpose, the man interpreted the coincidence as being teleological, that is, *caused by* the purpose it served. The man's bit of luck does seem, Aristotle notes, like the kind of thing that intelligence would have caused. But Aristotle insists that accidental functionality (originality in my terms) should not be confused with determined functionality (directionality in my terms). He did not suppose, as we might, that the former can eventually result in the latter.

ARISTOTLE CHRISTIANIZED

Somewhat unjustly then, Aristotle's teleology is remembered for anthropomorphizing natural processes. This mis-association is due, in part, to the co-opting of Aristotle's teleology by Christian mentalist teleologists.

13th century theologian Thomas Aquinas (1225-1274), in particular, was instrumental in transforming Aristotelian non-mental teleology into Christian mental teleology. Aristotle argued that the finality observed in the world leads to the conclusion that there must be intrinsic guiding principles that result in self-organization. Aristotle's *telos* is non-material, but should not be understood, therefore, as spiritual or supernatural. One might think of a non-mental telic principle as a range of possibilities that limit and direct development, maintain order, and balance proportions. Based on this, it is possible to make the argument that what we now refer to as self-organization was simply known to Aristotle as directional final cause. Granted, this view of Aristotle's teleology is made possible by the hindsight supplied by contemporary physics. Yet it makes better sense of Aristotelian teleology than does the Christian interpretation.

Self-organization in nature led Thomas to posit not a system of generative rules that emerge over time (as contemporary sciences do) but an *internal nature* that guides processes and is (unlike Aristotle's) placed there and controlled from the outside by God. In this view, man's—or any natural system's—internal nature is imposed by an external agent. In *Summa Theologica*, Thomas makes the very famous and influential analogy between human intentionality and an archer shooting at a target,

> *The natural necessity inherent in things that are determined to one effect is impressed on them by the Divine power which directs them to their end, just as the necessity which directs the arrow to the target is impressed on it by the archer, and does not come from the arrow itself. There is this difference, however, that what creatures receive from God is their nature, whereas the direction imparted by man to natural things beyond what is natural to them is a kind of violence. Hence, as the forced necessity of the arrow shows the direction intended by the archer, so the natural determinism of creatures is a sign of the government of Divine Providence.* (I:103:1 ad 3um)

Thomas makes the purposeful being a tool of God. This kind of abuse of final causes eventually inspired critiques from Francis Bacon (1561-1626), René Descartes (1596-1650) and Spinoza (1632-1677). It is a misinterpretation of teleology to make final cause into a divinely and *externally* determined nature. Final cause is better understood as an *intrinsically* determined set of relations (defined and altered by context), involving limiting factors, and resulting in order, proportion, and harmony. Thomas' powerful analogy between archery and goal-directed behavior separated the self from the goal or target. The self-organized interrelation that would have made archer, bow, arrow and target all part of one organic system becomes impossible under this metaphoric scheme and necessitates an external final cause. Sometimes analogies are more confusing than helpful.

The tool metaphor is damaging to teleology. The organ metaphor works better. My friend neuroscientist Walter J. Freeman[3] believes that Thomas' work in biology is much less insistent upon the idea of an externally imposed nature than is his theology. Freeman has revised Thomistic intentionality in his book, *How Brains Make up Their Minds,* arguing that brains deal with "meaning" not bits of information. Although in general, the uses put to Thomas' teleology align him with old-school mentalism, Freeman has done otherwise. Many a teleogist's work may be worth salvaging, and Freeman has found Thomas work useful in biology, where the organ rather than the tool is the source of the prevailing schema.

BACON AND KANT ON ANALOGIES AND AFFINITIES

"Final causes," writes Francis Bacon, "have relation entirely to human nature rather than to the universe, and have thus corrupted philosophy to an extraordinary degree" (59). Late in his life, a much-perturbed Bacon attempted to distinguish between affinities and analogies and thereby to eliminate mentalistic teleology from science. In Book II Aphorism 27 of *Novum Organum*, Bacon proposes that science should be directed toward,

> *investigating and noting the similarities and analogies of things.... But there is here a strict and serious caution to be observed, that we should only accept as Conforming and Proportionate Instances those that mark out physical similarities...that is, necessary and essential ones, grounded in Nature, not contingent and apparent.* (194)

Using methods of inquiry similar to those used later by non-mental teleologists, structural evolutionists and complexity scientists, Bacon sought governing principles intrinsic to the physical system. Bacon recognized "a similar nature" between "the structure of the ear and places that return an echo [*e.g.*, a cave]," between "the roots and branches of plants," and between "teeth in animals and beaks in birds" (193). Prefiguring 20th century structuralist D'Arcy Thompson, Bacon argued that the study of their similarities might provide general knowledge regarding the physical laws of their construction.

---

3. Walter Freeman and Jennifer R. Hosek received a Dactyl Foundation award in 2006 for their essay, "Osmetic Ontogenesis, or Olfaction becomes You."

Bacon rejected the scientific value of teleology because he had inherited a thoroughly mentalistic account through the Christian church, which relied heavily on spurious analogy. But his own approach actually comes close to non-mental teleology.

In "Critique of the Teleological Judgment" (1790) Immanuel Kant attempts to reinstate *non-mental* teleology by making a clear distinction between affinity and analogy. He insists that there is only a speculative analogy between the organization and functionality of human artifacts and organization and functionality of natural processes. Therefore, he argues, one cannot infer intention in nature as one might infer intention in another human being. There may be an affinity between the minds of two human beings, but there is only an analogy between human purpose and cosmic *telos*. Nevertheless, Kant maintained that the analogy between art and nature was useful as a heuristic device for understanding organic processes and self-organization—as long as the important distinction was maintained that these processes were determined by a cause intrinsic to the system itself while human artifacts were determined by actual external agents who acted on the object from the outside (I would say Kant didn't understand how art/invention actually gets done, but that's a finer point I don't need to make here). Kant's non-mentalism may be compared to Aristotle's in this regard: a conscious agent responsible for conceiving of the telic principle is not necessarily implied by the principle itself. And also like Aristotle, who had made a distinction between accidental functionality and determined functionality, Kant made a distinction between *extrinsic physical ends* (*e.g.*, rain happens to be good for plants) and *intrinsic physical ends* (*e.g.*, a tree grows by means of its own photosynthesis and is both cause and effect of itself) that demonstrate recursive causality.

CONCLUSION

I have made this tour of history in these two last chapters as brief as possible. In Part III where I discuss different types of teleology, I will add more references as needed, but in the end, my account will still be limited to those teleologists whom I have found most useful in distinguishing directionality and originality, non-mentalism and mentalism, affinity and analogy, course-affirming stochastic resonances and course-changing stochastic resonances. I am less concerned about covering everything and being a good academic, which too often means being a pedant, than I am about making my point clear and brief.

Some may note the conspicuous absence of Friedrich Hegel in my short history of teleology. When I looked into Hegel I found him useless for my purposes, largely because in his sense of directionality there is a well-defined goal of unification. Directionality in my sense is a trend that might go any way originality might take it. Moreover, teleological function creates *diversity* as organisms adapt to various dynamic contexts, as many biologists have noted, not *unification.* Hegel, and others like him, are associated with "Grand Narrative" teleology rightly condemned by postmodernists like my friend Arkady Plotnitsky. I say Hegel is no true teleologist without an understanding of directionality *and* originality.

Telic phenomena have two aspects: directionality and originality. Teleology has been misinterpreted as the study of an *external* intentional force guiding development, this largely to the association of mentalism with the role of chance (or chance analogical relationships) in originality. Teleology has also been misinterpreted as the study of *predetermined* principles that control development, this largely due to the role of constraints in directionality, which results in noticeable trends. If the two aspects are noted simultaneously then many misunderstandings may be avoided.

I also want to note here that I have been criticized by well-meaning friends for coining these terms "directionality" and "originality." Disparaging the use of neologisms in general, Murray Gell-Mann once complained that scientists would rather use each other's toothbrushes than each other's terms. I am not against using tools invented by others if they work well, but I haven't found suitable alternatives. I chose the terms because they are descriptive of the phenomena that are variously, and less clearly, labeled elsewhere. Directionality emerges fortuitously as a trend (but not toward anything) and originality fortuitously innovates.

With regard to these two terms, let me make here one brief comment on beauty, which may seem too brief in a book largely about artistry. Directionality leads to the kind of beauty that's easy to see and which involves structure, harmony, proportion, and balance. Originality leads to the kind of beauty that is more complicated and involves issues of meaning and association. I think only fully teleological art, having both directionality and originality, can be beautiful. A lover with a classic face isn't truly to you beautiful unless you're in love, and love, attachment, and affection grow out the positive associations that we make with our lover's features and gestures. So, to me "beauty," or that quality in art that is of aesthetic value, is cheapened if we think of it as mere prettiness. What is beautiful to me is teleological, *i.e.* that which has form *and* significance, directionality *and* originality.

Those friends who don't like my new terms have had trouble remembering the definitions I give to each term and often, for reasons I do not understand, confuse the two. So, although I have already defined them numerous times, I do so again, and will again, each time in a slightly different way, so that eventually, I hope, my intended meaning will emerge in the reader's mind.

Teleologists focusing on the aspect of directionality have been interested in archetypes, self-maintenance, self-organization, species coherence, and homeostasis, that is, emergent dynamical *stability*. Teleologists focusing on originality have been interested in functional adaptations, species evolution, accidental functionality, or fortunate *change*. These two aspects, emergent lawfulness and adaptability, make natural systems telic, that is, progressive or creatively organized. In this view, only when activity involves both directionality and originality can it be called intentional or, in my view, artistic. With these considerations in mind, I now reinvestigate teleology's influence on narrative aesthetics.

PART III

APPLICATION: THE INFLUENCE OF TELEOLOGY ON STORIES

Chapter 8

ARISTOTELIAN STORIES

We witness, you explain, not endgame but succession,
a process of soil and plant selection,
a gathering of just-so elements under magnificence of canopy,
crowning in glory, protecting new growth, harboring
the many and the whole which promulgates with
each light breeze a thing unto itself, a climax forest.

–Kevin O'Sullivan

In addition to the broad categorization I make between mental and non-mental teleologies, a number of further refinements can also be made. We can say that Aristotelian teleology, which I've already begun to discuss, is old-school non-mentalism. What I call "analogical determinism" is old-school mentalism. The term "deterministic fortuity" is the name I give to a newer and improved version of non-mental teleology. Pragmatic teleology is a much-improved version of mentalism. This is by no means an exhaustive list. Possibilities of finer distinctions are endless, I suppose. I only want to mention the general types that I've noticed again and again, especially in literature. One's conception of teleology has a profound effect on the way one organizes a story.

The fact that Aristotle deals with teleology in both *Physics* and *Poetics* gives us some indication of a natural relationship between one's theories of causality and aesthetic preferences. In *Physics*, Aristotle confines himself almost exclusively to the consideration of only one part of the twofold mechanism that creates teleological behavior. He was concerned with directionality and the maintenance of order. As mentioned in regard to Aristotle's anecdote about the man on the way to the market to get money who, luckily, ran into someone who owed him money, Aristotle did not approve of those who liked to over-interpret the role *accidental functionalities* in causality. In *Poetics*, Aristotle refers to the story of a man who goes to festival and is killed by a falling statue of Mitys as he is looking at it. This event initially seems like pure accident; however, as it turns out, the man was Mitys's murderer. By reading this tale backward in time, the apparent accident seems to be "caused," as narrative theorist Wallace Martin puts it, "by the logic of the ethical dimension" (128). In *Physics,* Aristotle had already made it clear that the lucky usefulness of such events seem like the sort of thing that results from intelligent deliberation because it serves a purpose (Bk. 2 Pt. 5). But he advises, accidental functions are not to be regarded as purposeful.

If a playwright chose to include a fortuitous coincidence, its only virtue would be that it seemed to "have an air of design" (Poetics Sec. 1 Pt. 4), and Aristotle was aesthetically pleased with what he considered authentic teleological design. Aristotle insinuated that unsophisticated audiences that might be awed by tales like that of Mitys because they suggest the intrusive and unrealistic intervention of gods. But as Aristotle argues,

> *the unraveling of the plot, no less than the complication, must arise out of the plot itself, it must not be brought about by the Deus ex Machina.* (Sec. 2 Pt. 15)

In *Oedipus Rex*, a play by Sophocles that Aristotle admired, Oedipus's father, king Laius, learns from an oracle that he is destined to be killed by his own son. Lauis tries to murder Oedipus, but events prevent it, and Oedipus grows up unaware that his adopted parents are not his real parents. When an oracle tells Oedipus that he will one day murder his father and marry his mother, he leaves home to prevent such things from occurring. Running to escape his fate, he meets Laius at a crossroad and, in an incident we would call road rage, he kills his biological father. Various incidents lead Oedipus back to his original home, where he is made king, unbeknownst to him replacing his own father whom he has just killed, and where he is married to the king's widow, Jocasta, his own mother. It may be that because these extraordinary coincidences have already occurred when the action of the play begins that they do not offend Aristotle's aesthetics too much. The actual events in Sophocles's play concern only Oedipus' search for the truth about what he has done. Once he learns everything, ashamed he blinds himself as punishment for not having "seen" the truth.

Although this story involves incredible coincidences, it is true that it is Laius's and Oedipus's reactions to prophecies that help them come true.

Prophecies in ancient Greek narratives tend to be self-fulfilling. It is the men themselves, not gods in fiery chariots, who complicate the plot. In *Poetics,* Aristotle allowed that, with stories, the "probable impossibility" should be preferred to the implausible but possible (Sec. 3 Pt. 25). Aristotle's sense of the concept of "probable" likely entailed an archaic meaning of "respectable" or morally sensible, which survived into the early modern English version of the term (see Hacking 19), so we might assume that Aristotle's sense of "probable" refers to outcomes that are poetically just, even if the set-up is physically improbable.

Personally, I have never found Aristotle's *Poetics* to be overly restrictive, although many authors and critics have railed against his proscriptions against sub-plots, against including anything that doesn't further the action in some way. But I'd like to point out that from Oedipus' perspective at the beginning of the play, the events of his life *are* random. Initially, he isn't aware of the relationships that connect him with Laius and Jocasta. That's pretty much what's interesting about the story. In *Poetics,* Aristotle emphasizes the importance of how the protagonist comes to know what has previously been hidden. The technical term *anagnorisis* describes a point of discovery that leads to a reversal of fortune, catastrophic in tragedy, happily resolved in comedy. Following this recognition is a sharp change in the direction of events, known as *peripeteia.* The change is the result of a contextual shift that makes a new reading of previously known facts possible. Oedipus's questions and investigations add up to the truth about who he is and what he has done, revealing the meaning of the coincidences. One cannot say that Oedipus's story is linear, in the sense of tracing mechanistic cause and effect only. It turns upon a reorganization of facts.

What Aristotle's aesthetics condemn is the lack of organization and unity, such as we find, it turns out, in the speech of people with the formal thought disorder schizophrenia, which we may define as a non-teleological mental disposition (with bi-polar disorder being a teleological disposition). Although I am no authority on psychology, I can relate a thought passed on to me by James Goss (at the time) a graduate student in Psychology at the University of Chicago. He wrote to me about one of my papers called, "The Poetics of Purpose" because he had recently published a paper called, "The Poetics of Bi-Polar Disorder," which, it so happens, makes similar assertions about self-organization and teleological thinking. Goss' thesis is fitting in this discussion of Aristotle's *Poetics* in more ways than overlapping titles. According to a well-known study in the field, some of the symptoms of schizophrenia include "inability to sustain *goal-directed* conversation" (emphasis mine), "speech that is significantly delayed in getting to the point and is characterized by excessive detail and parenthetical commentary" (I'm a little guilty here, I admit), and a disregard for "conversational conventions," which "is often irritating and offensive to the listener" (Sanfillipo 403). All these, we may say, are symptoms of some postmodern literature too, which specifically rejects Aristotle's aesthetics.

Of course, we all feel there is a fine line between art and madness, but artistic madness these days is specifically associated with bi-polar disorder (see Jamison), the symptoms of which are more than reminiscent of the self-organizing formalisms I've described, involving stochastic resonances, and include what are called, "clang associations" or "strings of words chosen because they are similar in sound and often rhyme." The bi-polar stage between depression and mania, often involves a high level of linguistic competence, "shifting back and forth between related topics," in a "relatively coherent and goal-directed" way," though sometimes focusing on what some people may consider "too many levels or aspects … simultaneously" (Sanfillipo 403). Bi-polar speech is a little too "artful," for general audiences, in other words. The bi-polar manic state is characterized by play between "multiple semantic levels," "sound and meaning in puns," and "ambiguous referents that allow [the speaker] to move in-between contextual backgrounds" and "maintain goal-directed discourse" in surprising or sometimes too surprising ways by destabilizing reference points. One might say that bi-polar mania is hyper-teleological discourse.

If bi-polar tendencies are associated with artistic temperament because teleological, then it cannot be said Aristotle's teleology constitutes a proscription against artful stories. Oedipus' final act of blinding himself metaphorically represents his not having understood—seen clearly—the truth of his actions. This is an artful—or bi-polar if you will—type of interpretive response. In many ways, in giving proscriptions, Aristotle merely gives good practical advice to playwrights, who worked within specific constraints. The plots he likes may be a little too tight for my personal tastes, which lie somewhere between Kafka and Pynchon, but I have little argument in general with his aesthetics, which seem fairly flexible insofar they don't forbid the use of chance occurrences in the plot.

Aristotle has been very useful to teleologists and narrative theorists who want to know *why* a story develops one way and not another. Like him or hate him, he will not be ignored. We may say that Aristotle set the stage for the four major teleologies and corresponding narrative structures that I look at here and in the next three chapters. He sketched out the possibilities of teleology in his definitions of both what he thought it is (non-mentalism) and what he thought it is not (mentalism).

Chapter 9

ANALOGICAL DETERMINISM

Thoughts create a new heaven, a new firmament, a new source of energy, from which new arts flow.

–Paracelsus

The teleology that developed during the dark ages did not suppose, as Aristotle had, that only regular events could be purposeful. Medieval Christians instead embraced the notion that *telos* causes especially fortunate or unfortunate accidents. According to this teleology, future effects can determine present accidents: Mitys's statue falls *in order* to punish his murderer. This all makes perfect sense if there is a supernatural author controlling events who experiences the past, present, and future all at once. Such an author would not be confined to linear causality, and a temporal end could affect the beginning. Time exists simultaneously to God the creator. It is this "fact" about religion that makes an analogical-deterministic teleology nonlinear and therefore conceivable. Anyone who has criticized religious teleology for being linear and end-directed has not understood this conception of divine temporality.

Although there is supposed to be no such thing as chance in this teleology—everything is purposeful, thoughtfully composed; nothing is merely random—human beings, existing in linear time can't comprehend God's nonlinear order and sense. Causality can go any direction in time and events and things not physically connected can be connected in God's mind. To humans everything that is really divinely ordered just appears random. They have little power to make predictions in this world ruled by this supernatural causality. Anyone with knowledge of the example of Abraham's attempted sacrifice of Isaac would have been powerless to predict that God would sacrifice his own son. However, this is exactly the kind of causality that exists in a timeless universe that turns on divine analogies. Only prophets and literary critics are in the position to guess how things might turn out. Ordinary folk generally can only recognize the prophetic nature of events with hindsight.

I have argued previously that this way of thinking does not represent a true teleology because purposes cannot be defined by someone or something external to the sphere of action. God is somehow beyond and above spacetime. As an intervener, He is an efficient cause not a final cause. Purposes ought to be self-defined. Teleology—I'll say it again—always involves inherent self-organizing processes. Even so, this strange and wonderful pseudo-teleology, which involves what I call *analogical determinism*, is indispensable to understanding what true teleology is. It's a primitive, half-ignorant, superstitious form of belief that nevertheless grasps in a profoundly accurate way what certain kinds of telic order *feel* like. In previous chapters, I have discussed how inanimate processes—for example chemical reactions—can involve "interpretations" in the sense of the parts interacting with a whole by means of semiotic causality. And so I have argued that interpretation and subjectivity are important parts of teleology, but they need to be "naturalized" and discussed scientifically. The teleology discussed in this chapter involves human interpretation, which can be ungrounded. Here I focus on originality, and subjective mentalism, which are absent in Aristotle's version. Unlike Aristotle's version, this teleology is quintessentially poetic, insofar as "poetic" refers not so much to formal qualities but the various interpretations formal qualities can elicit.

In the middle ages, divine speech was conspicuously audible in coincidences. Interpreting chance patterns in the "book of nature" became the new teleology. The belief that nature is a book—a work of art—has helped, partly, to establish our notions of art. Some aspects of art, we may say, evolved right out of a religious/superstitious view of reality. Medieval alchemist Paracelsus (1493-1541) illustrates well the widespread belief that the world could be read like poem. Nature, argues Paracelsus,

> *made liverwort and kidneywort with leaves in the shape of the parts she can cure.... Do not the leaves of the thistle prickle like needles? Thanks to this sign the art of magic discovered that there is no better herb against internal prickling.* (Qtd. in Hacking 42)

Paracelsus knew that mercury could cure syphilis. He reasoned, in what we would now consider perfectly Joycean style, that this is so because the name of the cure—"mercury," from "mercer," whence also "merchant"—was etymologically related to the place in which the disease was most often contracted, that is, in the "market place," so to speak.

This may seem crazy to us now, but later in history, this teleology gets revised and naturalized: a supernatural poet and punster will be replaced by selection processes (See Chapter Eleven on Pragmatism). Likewise the process of human authorship will also be naturalized: a homunculus-type essentialist self will be supplanted in explanations by a self-organizing selection process. These selection processes, as I argue throughout this book, are natural semiotic processes that involve "interpretation" or responses to signs. "Interpretive" responses can be found even in inanimate systems where there are parts functioning as indices of holistic states. Therefore much of what we may learn about the interpretive processes of a teleology that is determined by analogies is relevant to later, revised teleologies, and can help us understand artistic and natural processes that involve proto-interpretation, iteration and feedback.

Although a human author does not actually experience narrative past, present, and future in an instant, he or she can recall these more or less simultaneously, and most authors rethink and revise, and so in the creative process, later events do not necessarily follow earlier events in time. So although analogical determinism may be a superstitious teleology with regard to natural history, it is appropriate with regard to art. In stories, details and events unconnected in space and time do get linked mentally. Readers and viewers in their turns make the linkages too and experience the teleological character of this kind of art.

EVERYTHING WORKS OUT FOR THE BEST

Many twentieth-century novelists reject the happy ending because it is too much associated with Providential views of the world. In life sometimes things work out for the best; sometimes they don't. According to a twenty-first century view of teleology, "best" would defined by context, and what works out may not necessarily be best for the all parties interested. What is assured is that something will work out and that something will be, in some sense, best for whatever flourishes thereby.

Providential teleology may be more of a result of politics and economics than religion, philosophy or science. Medieval Christians were, presumably, consoled by the idea that everything is just and intelligible, even though it does not appear to be so. Tragic accidents and catastrophic events were common amid the political instability and poor health conditions of the Middle Ages. Adopting a stoic attitude, believing that behind the apparent disorder is order, might have helped some people endure all sorts of injustices and abuses (and might have helped other people continue to inflict them as well).

Medieval author Boethius[1] takes up the stoic attitude in his book *Consolation of Philosophy* (524), which he wrote in prison while awaiting horrible torture and death for the crime of being on the wrong side of politics. As he describes himself, Boethius was actually an altruistic man of the senate. When he was unfairly sentenced to death, he could not understand the apparent injustice that had befallen him. He consoled himself with the philosophy of stoicism that taught him how to rationalize all apparent evil. Stoicism showed how everything ultimately serves divine purpose. In the end, he welcomed his death because he believed it would fulfill divine Providence, somehow.

Geoffrey Chaucer started writing the *Canterbury Tales* in the 14th century shortly after translating Boethius's *Consolation of Philosophy*. The narrative structure of the "Knight's Tale" reflects the notion that apparently random events are actually caused intentionally, by a divine being, in order to achieve some end. If every event serves some ultimate purpose, then the poet who is trying to construct a narrative might have some difficulty deciding what is crucial to the story.

The Knight begins his tale with a digression. He mentions the mythological Theseus and his new bride Hippolyta. Then he explains how on his way home from his wedding, Theseus meets some women who have lost their husbands at the hand of a tyrant named Creon. It is then related how Theseus goes to fight Creon, and after battle, Theseus stumbles upon a couple of injured knights, and it is these knights who are the real protagonists of the story, which finally begins.

Unlike in an Aristotelian narrative, which would have begun with the knights, in this narrative, anything and everything may be causally connected in God's ultimate plot. This is perhaps why the narrator has difficultly knowing what to include, what to leave out, and where to begin. The knight's descriptions end up being too detailed and his tale is too crowded with

---

1. Boethius was not a Christian, but his Stoicism later became very much a part of Christianity.

characters. Some characters turn out to be instrumental to the plot. Others do not. The plot is driven, for the most part, by chance meetings and instances of luck. The riders of directionality are constantly switching horses. One of the knights (in the story) claims it is pointless to say whether some event comes about "by accident or destiny, / For as events are shaped they have to be" (58). Providence is thus indistinguishable from chance *and* determinism. The knight, like Boethius, is a stoic. His tale is concerned to show that ultimately all suffering is preordained and serves a divine plan. Unlike Boethius's story, the "Knight's Tale" should be seen, I think, as rather funny.

In 1759 Voltaire wrote *Candide* as a comic critique of this extreme brand of "everything works out for the best" teleology. In Voltaire's tale, Dr. Pangloss (whose name means "interprets all") attempts to rationalize even the most awful tragedies. He claims all events—from gratuitous murders to natural disasters—ultimately serve divine purpose. This kind of *telos* has a very obvious mental aspect. Analogical determinism can adapt to any particulars; it can distort, manipulate, and reinterpret virtually anything in order to make it fit an idea.

OUTSIDE OF TIME

It is St. Augustine who is principally responsible for the argument that God is outside of linear time, which allows eternity to exist all at once for Him. According to medieval Christian doctrine, codified primarily by Augustine in the fourth and fifth centuries, Providence is initially unintelligible only because human beings and God experience time in these different ways (see *Confessions* 11.11.13). In the end of the linear experience of time, the meaning of everything will become clear for humans.

Augustine's theological doctrine provided the frame with which he could make sense of apparently meaningless information. Under the pressure of Augustine's reading, the Old Testament predicts, and authenticates thereby, Christ's incarnation and the Church's power. Every sacrifice described in the Old Testament is interpreted as a foreshadowing of Christ's sacrifice. Physically unrelated events (*e.g.*, various sacrificial offerings) are linked by virtue of the *idea* they reflect. Literary critic Erich Auerbach calls this kind of pattern a "vertically linked figure,"

[A] connection is established between two events, which are linked neither temporally nor causally—a connection, which it is impossible to establish by reason in the horizontal [empirical] dimension ... It can be established only if both occurrences are vertically linked to Divine Providence, which alone is able to devise such a plan of history and supply the key to its understanding ... This conception of history is magnificent in its homogeneity ... earthly relations of place, time, and cause had ceased to matter, as soon as a vertical connection, ascending from all that happens, converging in God, alone became significant. (74ff)

The homogeneity noted by Auerbach is a special characteristic of analogical determinism. Each part of space and time is integrally related to every other part, so that each can be understood as a "sign" of the whole, a collapsed infinite point of simultaneity. If the reader knows the whole, the principle for which everything exists, then he or she can read the meaning of every sign, and every sign says essentially the same thing.

Augustine was a rhetorician and semiotician in addition to being a church father, as well as a person with a colorful past, who kept a concubine, "The One," for 13 years, left her to become engaged to a girl not yet of marrying age, took a second concubine to hold him over until he could marry, and finally gave all up for the priesthood. He spent the rest his life grieving over and recounting (not without some pleasure, perhaps) his past sexual intimacies. Before converting to Christianity, which he in fact helped to create, he was intensely involved in Manichaeism, a kind of religious dualism that promoted the belief in equal and opposed forces of good and evil. It purported to be a religion of knowledge (Gnosticism) opposed to the Christian religion of faith. Augustine later wrote many refutations of Manichaeism and also "specialized" in discriminating between superstitions and belief in "real" miracles. He was, to my mind, a wonderful poet too. As a young person, I thoroughly dog-eared my paperback copy of his *Confessions*. His description of the relationship between heaven and God I found extraordinary and stunning, if unsettling. He likens heaven to the striving human creature who is not "co-eternal" with God, and can never "shed its mutability." "But," he continues addressing God with the insane hope of a romantic lover to his Laura,

being always in your presence and clinging to you with all its love, it has no future to anticipate and no past to remember, and thus it persists without change and does not diverge into past and future time. How happy must this creature be, if such it is, constantly intent upon your beatitude, forever possessed by you, forever bathed in your light. (287)

No doubt Augustine's influence is there in my critique of romantic love I mentioned earlier with regard to the "Gottlieb" theme in my own fiction.

In *City of God*, Augustine—who must surely have been influenced by Talmudical Hermeneutics, Jewish interpretive practices already well-established—had a way of shaming the purely literal-minded reader, whose failure to look for divine meanings he considered "folly and insanity" (106), "mad impiety" (183) and "mad contention" (240). Although none can easily guess how the future will fulfill God's intentions, he thought it was a Christian's duty to attempt to decipher God's will as signified in the *book* of nature and in historical events, as well as in the Bible. Augustine insisted on both the literal *and* the figurative readings of any material fact. He points out that:

the Scripture itself, even when, in treating in order of [sic] the kings and of their deeds and the events of their reigns, it seems to be occupied in narrating as with historical diligence the affairs transacted, will be found, if the things handled by it are considered with the aid of the Spirit of God, either more, or certainly not less, intent on foretelling things to come than on relating things past. (569)

Resemblances and poetic patterns in history, he claimed, provided the clues necessary to interpret divine will that might be encoded in apparently meaningless details:

For, in the manner of prophecy, figurative and literal expression are mingled, so that a serious mind may, by useful and salutary effort, reach the spiritual sense... (744)

After Augustine, many early theologians relied heavily on verbal, visual, or situational coincidences, as they believed in the tradition that everything in a sacred text had to be intentional. Nothing could be the mere product of historical causes. If they were not immediately struck by a sense of suggestiveness, the interpreters actively forced textual elements into interpretable patterns. "Seemingly petty details," writes Biblical scholar James Kugel, were especially tempting for interpreters:

the names of unfamiliar persons or places, narratives that seemed to have no overriding theme or message, or laws that were entirely too occupied with mundane matters. These cried out ... for some additional, overarching significance or simply seemed to suggest, in view of both their curious details and the lofty provenance attributed to them, that some other meaning beyond the obvious one had been the author's intention. (81)

Arguing along similar lines, literary critic Frank Kermode writes,

Interpretation abhors the random, which is one reason why, in the most modern school of criticism, it has become a dirty word, a term of censure. Interpretation seeks relations.... It will find, in some secondary world of magic and ritual, an explanation for the lucky.... It may go on to provide this fiction with mythical structure, a satisfying spiritual order, instead of the trivial carnal order of the primary narrative. (10)

Kermode's term for the kinds of patterns operating in analogical determinism is "occult structure."[2] He stresses the extent to which such interpretive practices were fully established in the early Hebraic tradition and were extremely influential for the Gospel writers. He found that St. Mark, in particular, strove to assign meaning to certain facts that other saints had left uninterpreted.

For example, he notes that Mark dovetails two unrelated stories, the first involving a young girl known as Jairus' daughter, who is raised from her death bed, and the second involving a woman who is cured of a hemorrhage when she touches Jesus' robe (5:21-43). These two events are only incidentally related: while Jesus is on his way to Jairus house, he happens to pass the woman who touches his robe.

In *Mark* the age of the girl is given as twelve. None of the other Gospels mentions her age. Mark also indicates that the woman with the hemorrhage had been ill for *twelve* years. According to Kermode, this "chiming" has been dismissed as a coincidence by scholars who have insisted, "that's just the way it really was" or who believe that the word for "twelve" is simply a vague denotation meaning a dozen or so. In this way, the pattern can be explained away. But Kermode objects to such dismissals:

2. My term for them is "phenomenal patterns." See Alexander, "Poland and Polonius, a Coincidence?"

> *in matters of this kind there is really no such thing as nonsignificant coincidence, and we are entitled to consider that this coincidence signifies a narrative relation of some kind between the woman and the girl.* (132)

Kermode guesses that Mark is "saying something about sexuality" by opposing menstruation with pre-pubescence, but he adds that his task is "not so much to offer interpretations as to speak of their modes, their possibilities, and their disappointments" (133). Here Kermode has identified a special property of poetic patterning, which, because of its dubiety, functions as a portal to another level of meaning outside the text. It is an example of telic originality. Kermode's interpretation is not verifiable, but as he himself notes, this is what makes it interesting. With analogically determined teleology one can only hope to guess at the correct meaning. Unless one can directly ask the author, who may or may not know herself, empirical verification will likely remain elusive.

Another interesting aspect of Kermode's analysis of the *Gospel According to Mark* is the way he treats it as a work of literature. Interpreting God's meaning has morphed into interpreting a human author's meaning. When the institutionalized study of the novel began to be established in universities around 1900, it adopted the interpretive practices that had been used by theologians, Biblical scholars, and alchemists because, not coincidentally, they are appropriate to the study of novels.

In "Prophecy" (1927), British novelist E.M. Forster engages in literary criticism influenced by these religious attitudes toward nature and history, describing a coincidental pattern in *Moby Dick,* which, he claims, gives the work "prophetic" power. Like Kugel and Kermode, Forster notes that it is the apparently meaningless coincidence that can be most suggestive of a higher level of intention.

At the beginning of the novel, Melville includes a long sermon, during which the preacher opines: "Delight is to him whose strong arms yet support him when the ship of this base treacherous world has gone down beneath him." Forster argues that,

> it is not a coincidence *that the last ship we encounter ... before the final catastrophe should be called the* Delight; *a vessel of ill omen who has herself ... been shattered by him* [Moby Dick]. *But what the connection was in*

> *the prophet's mind I cannot say, nor could he tell us. Immediately after the sermon, Ishmael makes a passionate alliance with ... Queequeg ... Towards the end he falls ill and a coffin is made for him, which he does not occupy, as he recovers. It is this coffin, serving as a life-buoy, that saves Ishmael...., and* this again is no coincidence, but an unformulated connection that sprang up in Melville's mind.... *It is wrong to turn the* Delight *or the coffin into symbols, because even if the symbolism is correct, it silences the book.* (140, emphasis mine).

The "non" coincidence here Forster recognizes as being indicative of the author's intentions. Forster resists assigning a symbolic significance to the coincidence, which would be to argue that the pattern is directional or conventional; however, he implies some significant association between destruction, salvation, delight, and homosexuality. Forster cannot prove the argument because the association is a singular event, a case of originality. If one looked elsewhere in Melville's work and found the same pattern of associations recurring, only then could one begin to make an argument that the pattern has relative objectivity as an intentional pattern that is particular to its author, a symbol. However, Forster's task is, like Kermode's, not to offer *the* correct interpretation, which may come with directionality, so much as to explore interpretive modes, possibilities, and disappointments of the telic aspect of originality.

In *Recent Theories of Narrative,* Wallace Martin proposes the term "odd textual conjunctions" for suggestive coincidental patterns. He realizes that "to make such connections is to presuppose a purpose," but he stresses, as I do in this chapter, that he is justified in doing so because "by accident and design the text was produced by a writer" (123) who is outside of narrative time. In some cases, Martin is also aware that he may very well be reading too much into random details. "At times we may wonder if odd textual conjunctions were planned" (123). But as we read, we think we see hints of the future, and, as we reflect back on the work, we are able to distinguish between real clues and false ones. As Martin remarks,

> *Assumptions about causality lead to conjectures about the future; ... We read events forward (the beginning will cause the end) and meaning backward (the end, once known, causes us to identify its beginning).* (127)

This process, of constantly looking both forward and backward while reading, dubbed "double reading" by narrative theorist Jonathan Culler, is related to the God-like stance beyond time which Augustine sought to approach with his double reading of the Bible. Likewise, the practice of reading a novel forward and backward helps recreate the author's experience of poetic simultaneity.

In his study of structural poetics, Tzvetan Todorov suggests that "gnoseological" is the term to describe narratives that are more concerned with the purpose or meaning of the events than how events happen. In a gnoseological narrative, coherence is poetic, not historical, and it is based on resemblances. According to Todorov, coincidences that do not advance the plot seem to signal intentional meaning. They function as "indices, and oblige us to set out on an interpretive track that is independent of the principal semantic line" (58). This reminds us again of Kugel and Kermode's argument that chance patterns seem to "cry out" for interpretation.

It could be said that Augustine's insistence on the prophetic possibilities of apparently meaningless facts in historical narratives encourage superstitious interpretive practices. This may be an inappropriate attitude to have toward historical events of which there is no Author, but it is not an inappropriate attitude to have toward a poetic narrative. A superstitious attitude toward a text makes an insightful interpreter. An analogically determined narrative does represent the way humans—animals with language—experience time, in an almost God-like way. When Hamlet (4.4) begins to feel that all "occasions"—various random events he witnesses and coincidental patterns he sees—are *meant* to "spur his dull revenge," he perceptively notes that this is because humans have the special capacity of language, "such large discourse," and are unique among animals in their ability to look "before and after" that makes them able to read events as if they were scenes in a play designed by an author outside of time (see Alexander "Poland"). This subjective experience is a fact of our existence. It is an objective part of our reality, a reality represented so well by poetry.

We who love poetry owe much to the superstitious attitude toward history and nature. Without it we would not be as practiced at guessing the possible meanings of those intriguing "mentalistic" patterns that point outside the text and may be indications of the author (lower case "a") who is also beyond the narrative frame experiencing the story's divine simultaneity.

WHAT CAUSES PATTERNS IN RANDOM EVENTS?

With few exceptions, believers generally find evidence for supernatural purpose in the improbable, good or bad luck, miracles, or even funny coincidences. When the ancients wanted practical advice for political matters, they looked to sources of randomness to tell them the will of the gods: they drew lots, examined animal guts, and listened to sibyls babble incoherently.

Coincidental regularities may be surprising to us, and thus seem to "say" something. But they can be entirely consistent with the laws of probability. Your odds of winning the lottery may be one in a million, but someone does win, about once in every million tries. Nevertheless, notes MacArthur prize-winning mathematician Persi Diaconis, "our intuitive grasp of the odds is far off. We are often surprised by things that turn out to be fairly likely occurrences" (Diaconis and Mosteller 854). That surprise, in turn, tends to encourage us to exaggerate the significance of the coincidence. Once a professional magician (he took to the road with a sleight-of-hand artist at age fourteen), Diaconis is often called upon to debunk apparent miracles performed by charlatans making use of our naturally poor sense of probabilities.

I met Diaconis at the Santa Fe Institute, and later, when I was in San Francisco, I drove down to Stanford University, to visit him in the Department of Statistics. In San Francisco, I had been meeting with a fellow Art and Science Laboratory member, Natalie Jeremijenko, a conceptual artist and clever prankster. She had recently devised an instrument that would detect the number of bodies that drop from the Golden Gate Bridge into the bay. These statistics the local police do not want publicized, as it tends to encourage more bodies to drop. When I went running in the park by the bay the next morning, I was confronted with sign after sign indicating things like, "Caution: High Cliff. Danger of Falling to Your Death" and "Do Not Jump," messages which seem to have the possibility of being more suggestive than prohibitive. It was the coldest, grayest, dreariest winter in August I have ever experienced, to echo Mark Twain, and the weather, together with the signs and Jeremijenko's device seemed "all occasions [to] inform against me" (*Hamlet* 4.4). I told Diaconis my story. It was sunny and bright in Stanford, and he provided the statistical explanation to dispel the unhappy interpretation of events.

A standard example of a "surprising" regularity is epitomized by the so-called birthday problem. In a group of twenty-three people, there is a fifty percent chance that two will share the same birthday. People tend to guess that the probability would be less than twenty percent based on the reasoning that twenty-three is less than twenty percent of 365, the number of days in a year. But calculating probabilities is more complicated and, as Diaconis points out, counterintuitive. What many people fail to consider when attempting to solve the birthday problem is that the matching birthday is *not specified beforehand*. The probability that two people in a group of twenty-three will share *any* birthday is much higher than the probability that two people will share, say, the specific birthday April 23rd.

A birthday match may not be a particularly interesting coincidence. When a coincidence is interesting, or useful, or appears just, it is more difficult to accept the meaningless probability of the event. As prolific 20th century writer and Catholic apologist G. K. Chesterton observes, apparently meaningful coincidences do happen, and they get our attention:

> *A few clouds in heaven do come together into the staring shape of one human eye. A tree does stand up in the landscape of a doubtful journey in the exact and elaborate shape of a note of interrogation. I have seen both these things myself within the last few days. Nelson does die in the instant of victory; and a man named Williams does quite accidentally murder a man named Williamson; it sounds like a sort of infanticide. In short, there is in life an element of elfin coincidence which people reckoning on the prosaic may perpetually miss.* (3-4)

Even though coincidental regularities may be easily explained away by the laws of probability, the temptation to believe that they are contrived (Chesterton calls these coincidences "miracles") is hard to resist, as is born out by this passage. Significantly, however, the coincidences mentioned here are, like the birthday match, not prespecified. That is, the speaker does not predict the particular shape that the clouds will form, nor does he predict the apparent infanticide. He only notes them after the fact. Any recognizable shape or coincidence will do, and the number is vast. Extending what we know from analyzing the birthday problem, we realize that the probability that *any* interesting resonance will occur on a given day to *any* person can be high. Nevertheless, when *you* are the one to which the coincidence is relevant, it is striking. As Chesterton emphasizes, "I have seen both these things myself within the last few days."

FREUDIAN NARRATIVES

A chance event that is said to occur *because of* its later effect is supposed to be supernaturally end-directed. The landscape presenting a doubtful tree in the Chesterton passage seems to express the speaker's emotions. Since the tree was there before the speaker arrived on the scene, its shape seems predetermined and fated for the accidental function it comes to serve.

Sigmund Freud (1856–1939) argued that we would be wrong to think that coincidentally useful events in the outer world are caused by cosmic purpose. He did think, however, that coincidentally useful or telling events in one's inner world *are* caused by the Unconscious, of which one may have no control or knowledge. Freud theorized that the Greek notion of fate was often a projection of these internal intentional mechanisms onto the external world. Unconscious actions express a person's hidden emotions by signifying a repressed state. For example, Lady Macbeth's compulsively washing her hands may indicate her feelings of guilt. Guilt here is assumed to be equal to feeling unclean. Even though this action may appear only coincidentally analogous to one's feelings, according to Freudian psychology, the telling act is actually being driven by the Unconscious. The Unconscious is purposeful, if you believe *it* does things for the sake of alleviating the pains of repression or expressing hidden desires.

In our lifetimes, Sigmund Freud's theories have gone in and out of favor. Nabokov famously rejected Freud's ill-fitting mythic masks and the psychoanalysis of literature, especially his own (see Boyd, *American Years 160*). While few psychologists today may diagnose patients with Oedipal complexes, many neuroscientists do recognize some Freudian concepts such as the unconscious and the phenomenon of repression.

Although my view of purpose involves the idea of self-organizing tendencies, which some might associate with Freud's Unconscious, we differ insofar as I understand that patterns form initially by formal constraints and not *because* of some function they may later serve. I would be inclined to say the Unconscious often follows neutral rules matching things that are merely near or merely like. From such a process of self-organization, a pattern may emerge that happens to fit with some other pattern. Whether or not that match may be considered significant vis-à-vis something else is another question. I find some of Freud's interpretations as dubious as Augustine's or Chesterton's. He, too, could not leave coincidental resemblances alone. He felt that chance patterns must in fact be directly caused for the sake of the function they serve or the meaning that they might have. The similarity between Freud's and Augustine's interpretive strategies is most obvious in a passage in which Freud argues that (what I consider coincidental) order found in arbitrarily chosen names or numbers revealed the "will" of the Unconscious.

In "Determinism—Chance—And Superstitious Beliefs" (from *Psychopathology of Everyday Life*), Freud gives an example, from his own experience, of the workings of the Unconscious. He argues that unconscious intentional order can be discovered in apparently random events. He relates the story of how, in a letter to a friend, he had written that he would not make any further changes to a particular manuscript, even if it contained "2,467 mistakes" (120). This was a number he had chosen completely at random. Later, however, upon reflection, he decided to look for an underlying principle to explain the choice of that particular number, because he believed, "There is nothing arbitrary or undetermined in the psychic life" (120). Once he had decided that the number might not have been arbitrarily chosen after all, he remembered that prior to writing the letter, he had been speaking to his wife about a man he had not seen in 17 or so years. He then remembered that he had been 24 at the time he had last seen the man. Therefore, Freud realized that he had underestimated; it had actually been 19 years since he had not seen him. Adding 24 to his then current age of 43, he came up with 67. Given that the number of mistakes that he had randomly specified as 2,467, Freud supposed his Unconscious (which was apparently better with figures) was expressing his annoyance that, in the time since he had seen this particular man, he had not accomplished much. This was his "mistake," he reasoned, thereby providing a direct "cause" for what had seemed to be an apparently randomly chosen number.

C. S. Peirce, Freud's older contemporary writing in 1878, explains how patterns drawn from arbitrary samples can be infinite if the rules are not decided beforehand. Freud allowed himself to use virtually any situation that might supply the number 2,467 with meaning. To illustrate how easily coincidental order may be found in random numbers, Peirce takes the first five poets listed in Wheeler's *Biographical Dictionary* and lists the age at which each died:

Aagard, 48.
Abeille, 70.
Abulola, 84.
Abunowas, 48.
Accords, 45.

Because Peirce allowed himself the freedom of finding any pattern at all among these random numbers, he was able to find a remarkable triad:

1. *The difference of the two digits composing the number, divided by three, leaves a remainder of* one.
2. *The first digit raised to the power indicated by the second, and divided by three, leaves a remainder of* one.
3. *The sum of the prime factors of each age, including one, is divisible by three.*

Peirce concludes: "It is easy to see that the number of accidental agreements of this sort would be quite endless" (176-177). Freud easily found accidental agreement between the number 2,467 and the idea of a "mistake" because he allowed himself to use any kind of rule. Instead of considering the number as a whole, he arbitrarily parsed it into 24 and 67, not, say, 2 and 467. Then he made 67 meaningful by *adding* his former age 24 to his current age 43. Any kind of manipulation (addition, subtraction, multiplication) to any combination of numbers would have been justified by Freud, as long as the answer it provided seemed to make a kind of sense. Similarly, Augustine claimed that any figurative interpretation of historical events was true so long as it fitted with Christian doctrine. Freud also allowed himself the freedom of drawing meaning from a circumstance (in this case, a conversation he had had with his wife) that was not related to the actual letter in which he had arbitrarily designated the number of mistakes as "2,467." In contrast to Peirce, who believed that accidental agreements were merely accidental, Freud concludes, "The analysis of chance numbers ... readily demonstrates the existence of highly organized thinking processes, of which consciousness has no knowledge" (123). Freud's argument that there is nothing random in the psychic life is ultimately based, in the example related above, on a mere and rather weak analogy between *a lack of accomplishments* and *a number of mistakes* in a manuscript.

Repressed persons will accidentally reveal their anxieties and divine justice will be served, but *how* the Unconscious or Providence may accomplish these objectives cannot be predicted according to empirical laws. Retrospectively, Freudians and Christian teleologists may link coincidental facts and events together, but accidental functions can only be described after an effect is produced, not before. Freudian narratives, then, belong to a category of narrative that has its roots in the medieval Christian Romance tradition.

LINEARITY

Literary critic J. Hillis Miller claims, in *Ariadne's Thread*, that a teleological principle operating in religious narratives can turn a sequence of events, a purely "metonymic line," into a causal chain, the logic of which is only apparent retrospectively. Ariadne from Greek mythology, we recall, was thrown into a labyrinth that had a beast at its center, and she found the only way out by following a thread. Writes Miller, echoing Derrida,

> *the line image ... tends to be logocentric, monological. The model of the line is a powerful part of traditional metaphysical terminology. ... Narrative event follows narrative event in a purely metonymic line, but the series tends to organize itself or to be organized in a causal chain. The chase has a beast in view. The end of the story is the retrospective revelation of the law of the whole. That law is an underlying "truth" that ties all together in an inevitable sequence revealing a hitherto hidden figure in the carpet.[3] The image of the line tends always to imply the norm of a single continuous unified structure determined by one external organizing principle. This principle holds the whole line together, gives it its law, controls its progressive extension, curving or straight, with some* arché, telos *or ground.* (18)

Miller's "metonymic line" refers to the simple metonymy of a picaresque, in which one event follows another without unified goal, a diarist's record of unconnected tasks completed. Miller notes that in a teleological narrative "the series tends to organize itself or to be organized in a causal chain." How do diurnal wanderings end up getting one somewhere—or forming some kind of pattern?

3. Miller is referencing a Henry James story here called, "The Figure in the Carpet." I will discuss this story in a later chapter.

I question the labyrinth as an appropriate metaphor for this kind of teleological narrative, though it is an ancient and popular one. Unlike Ariadne, who found the *only* way out of the labyrinth, a protagonist in a teleological narrative might take any number of routes to arrive at any satisfactory conclusion. The end is always a general *type* of resolution, not a particular one. A teleological Ariadne would not be in a labyrinth but among unnumbered stars, where if there is any "linearity" it would result from connecting one star with another forming a constellation in a way that is not predetermined. The dots in such a connect-the-dots game are not numbered, they do not follow prespecified rules.

I want to remind my readers that metonymy involves the substitution of one thing by something physically near it. The latter becomes a sign of the former. In a metonymic line, articles would be linked together by means of contiguity. For example, when Hamlet speaks of his mother's "incestuous sheets," he associates the linen with the act performed on them.[4] The sheets are not incestuous themselves but they are contiguous with the act and guilty by association: perhaps one might say they show signs of the otherwise hidden act. This is an example of *simple* metonymy, a substitution based on mere physical association via nearness.

In the case of synecdoche, a narrower type of metonymy, the contiguous thing standing in for something else is actually *part* of the thing signified. Synecdoche can be easier to understand than simple metonymy because the association between part and whole is more obvious than associations between two things that just happen to be near each other.

To put it into Peirce's terms, metonymy—simple or complex—would be an index of an object. The process of iteration can turn simple metonymy into synecdoche. One bird's "choreographed" movements are indicative (a synecdoche) of flocking. Birds that are contiguous with each other are signs to each other of the overall constraints; their limited interactions help build those constraints. The birds may start out being only randomly associated, but through exposure to each other they become more deeply related, eventually becoming part of a whole together.

4. Also Claudius is not Gertrude's brother but her brother-in-law. So Hamlet's uncle is really only like a brother to his mother. Thus, metaphor as well as metonymy is at work in this phrase.

A *chaining* together of metonymic signs, mentioned by Miller, might refer then to the kinds of self-organizing processes in which individual parts are trapped under the same constraints and interact such that they form a whole. But he seems to suppose that there is something unnatural going on when merely randomly interacting things suddenly organize into a whole.

The conception of teleology as "linear" is a subject this work has broached before. In *The Prophetic Moment*, Angus Fletcher notes that history in terms of a story with a well-defined plot involving creation, fall, redemption, and judgment is "somewhat misleadingly called 'linear.'" In this type of teleological narrative,

> *[s]eemingly chaotic and unrelated events are shown to have a progressive character; history appears to move in a certain direction. Because wandering bulks large in this story, the form of history in this tradition should be called "linear" only with the express understanding that with it the line is not a very straight line....*
>
> *... By showing that the wanderings of the chosen ones are momentously linked to the all-known but veiled design, the prophet "straightens" the twisting, labyrinthine shapes of profane time. When the children are lost, he unveils his prophetic gift, an inspired sense of direction.* (41-42)

What is key to the kind of teleological narrative that Miller has described is not its lawfulness, but its *arbitrariness*, which allows the prophet-interpreter to use *any* coincidental contiguity and/or coincidental similarity to make sense, sometimes showing connections in an apparently only random or metonymic sequence and sometimes making connections metaphorically.

If Miller's term "law" is applied to Freudian narratives, to Augustinian interpretation, to Chesteron's musings, or to any analogically determined teleological narrative, it must be done so *very* loosely. These laws turn on poetic associations; if they become lawful (directional) at all it would only be through frequent iteration. Miller feels that this type of teleology is oppressive, "monological." But what I see here is a wild freedom to interpret anything in virtually any way one wants. Metonymic *line* does not quite work for this concept, unless the line starts to form a messy and dynamic *web* of relations that form out of self-organizing processes, in which case we would need Arachne not Ariadne.

THE END

Retrospection is so important in analogical determinism. I want to end this chapter with some stories about endings. The first is about a painter, named Jim Gilroy, who was one of the first to show at Dactyl Foundation. This story is very analogically determined insofar as the temporal end seems to change the meaning of events that came before.

When he was thirteen, Gilroy stood in a crowd of onlookers one afternoon in midtown Manhattan and watched a man jump to his death from a skyscraper. He says he has never forgotten the sound, "like a car crash." Later as an adult, he happened to glance out his window and saw the dark brief streak of an upstairs neighbor, another suicide, dropping to his death. In 1999, Dactyl Foundation produced a documentary about Gilroy, his art and his relationships with childhood friends, who include the film director Larry Clark. The documentary is called *Don't Let Go*, and it opens on the balcony of Gilroy's high-rise apartment in TriBeCa in downtown New York. The camera pans from Gilroy and Clark over the edge and seems to jump down to the concrete below. *Don't Let Go* features Gilroy's reminiscences of his sneaking into the World Trade Center towers when they were under construction. He and his friends used to hang over the side of the observation deck, letting just the powerful wind hold them up. They did it for the adrenaline rush. One evening, one kid who was on drugs suddenly "just let go," as Gilroy recalls—with simultaneous dismay and understanding.

It was Gilroy who had to deliver the news to the boy's father, the most painful task he has ever had to do. Gilroy painted *Epiphenomenon* in 1999 with perhaps these events in mind. The painting does seem to be about the sensation of balancing precariously, and the figure makes an Icarus-like gesture against a gray sky. Later in the summer of 2001, Gilroy had obsessively started sketching people falling and jumping. Then on September 11th, he watched from his balcony in panic as people jumped from the towers to their deaths. One woman held down her skirt as she fell, and he had sketched that same sadly unnecessarily modest gesture days before. Does the World Trade Center tragedy change the meaning of Gilroy's artwork prior to September 11th? It does for Gilroy. The theme of jumping that had run through his life seemed to him to be leading up to this point in time. He felt singled out as witness whose experience of the "final scene," as it were, was fraught with disturbing complication and complexity.

Gilroy began working in abstraction thereafter because he had difficulty doing the figure. "The sky was so blue!" he keeps saying in disbelief, as if one might have expected the weather to have been gray in sympathetic response.

MEANWHILE

On September 11th, several blocks away, Frank Delessio was near the Brooklyn Bridge moments after the towers were brought down. On that fateful day, Delessio decided to scoop up a sample of 9/11 dust, and then he went to Tom Breidenbach's apartment in the East Village. Breidenbach—a poet, art critic, member of Dactyl Foundation, and one of Neil Grayson's biographers—carefully sealed Delessio's sample in a plastic bag (Jones). Nine years later, while reading a paper on the chemical analysis of 9/11 dust samples, I was surprised to see Breidenbach's name among the four people who had collected the dust samples tested by the authors of the paper (Harrit *et al.*). I was also surprised to discover that I know of another of the four people who collected samples, an artist neighbor of a friend (one of Neil Grayson's collectors) who lived on Liberty Street in the building directly across the street from the towers. Coincidence? Yes, of course, and as a novelist, I cannot resist narrativizing my non-fiction, which means subjectively tying together unrelated details into a theme.

Duncan Watts has offered the six-degrees-of-separation theory to explain away such connections. I first became familiar with the theory when Watts visited the Santa Fe Institute while I was there, and I also had dinner with Watts at a Columbia University event later, probably some time in 2001. Having no degree of separation between me and the man who worked on the theory about degrees of separation makes it seem somehow worth mentioning, again my novelistic impulses get the better of me. Though Watts was just an acquaintance, I knew Breidenbach pretty well and saw him often in the 1990s. When the name of someone you know well turns up in an obscure paper, it seems improbable, and you want to tell the story because coincidences are what make a string of facts into a story. "A funny thing happen to me on the way…." The "small world" theory, we shouldn't be too surprised, descends from a fiction writer, Frigyes Karinthy. In one of his stories, entitled "Chain-Links," published in 1929, the characters discuss how interconnected the world is becoming:

One of us suggested performing the following experiment to prove that the population of the Earth is closer together now than they have ever been before. We should select any person from the 1.5 billion inhabitants of the Earth—anyone, anywhere at all. He bet us that, using no more than five individuals, one of whom is a personal acquaintance, he could contact the selected individual using nothing except the network of personal acquaintances. ("Chain-Links")

Some may recognize Karinthy's ideas that were later popularized by Hollywood.

Given enough interaction and time, one would expect further connections to develop. The 9/11 plot for me became more interwoven several years later. The first American reporter to be murdered in Iraq after 9/11 was Steven Vincent, who, as it happens, was the first recipient of a Dactyl Foundation award in 1997 for his essay, "Listening to Pop," which explores the ways in which non-representational artworks acquire meanings, in an eerie sort of way. Formerly an art writer/editor with *Art & Auction* and the *Wall Street Journal,* after 9/11 Vincent found he couldn't write about art anymore, echoing both Jim Gilroy and Theodor Adorno's famous sentiment that poetry after Auschwitz had became inappropriate.

Incidentally, the 2001 recipient of the Dactyl Foundation award was Dominick LaCapra, who, having been chosen for the award months before, addressed audiences at Dactyl on September 28th, 2001 with a talk on his essay, "Trauma, Absence, Loss." In that essay, LaCapra notes that both Adorno and deconstructionists confuse real historical loss with metaphysical loss (people died, not a God), which results in a state of trauma, which the sufferer cannot work through. The pain of the loss can never be overcome, because the "lost" object never existed in the first place. Postmodern art, argues LaCapra, often represents reality as a state of irresolvable, unending trauma, a state in which "teleological" narratives are no longer possible.

Vincent was killed in 2005 shortly after the publication of a piece that he had written for the *New York Times,* questioning political practices of the reconstruction effort, which put too much control in the hands of fundamentalist groups. (Vincent had thought the point of the forces going into Iraq was to disempower fundamentalist terrorist groups.) In 2004, Vincent had published a memoir, *In the Red Zone: A Journey into the Soul of Iraq*, in which he writes about feeling destined in his search,

As we cleared the last curlicue of concertina wire, a silver orange rose over the horizon to break the pre-dawn symmetry of the landscape. I settled back in my seat, feeling my heart pound. This was the moment I'd been heading toward since the morning of September 11, 2001—and perhaps before that. I had quit my job as an art reporter and critic to do this, driven by conscience, support for the war, and some reawakened moral impulses I had yet to understand. (5)

When I knew Vincent, he was not obviously interested in politics. He was at that time more involved in questioning why some people, masochists, invited torture and pain. He had once confided in me that he did not understand it, and yet he did not want to dismiss it as simply bizarre. He wanted to understand it. I believe that these questions led him to become interested in what he would later refer to as the "death cult" of Islam.

When they found his body, some newspapers noted, around his neck was tied a red piece of material. Some implied that his killers had probably used it to strangle him. I knew that Vincent, a flamboyant dresser, often wore a red silk scarf, and I thought about the irony that his fashion choice came to be interpreted in this way. Apparently, he had been brutally beaten and shot, perhaps tortured. In view of the way his life ended, more painful irony now comes out in his writing,

What my journey would bring, whether it would answer my political and personal questions—or even if I would survive the damn thing—I did not know. (5)

According to the "small world" theory, every New Yorker ought to be related to Tom Breidenbach and Steven Vincent within at least six degrees of separation. There are probably many who were as surprised as I was to find themselves vicariously involved in some of the most traumatic events in U.S. history. Personal conspiracy theories such as Hamlet had—feeling that you are somehow destined to take action against what may be rotten in Denmark—are often founded upon the (to us, surprising but really) probable coincidences that weave our lives together like stories (see Silverstein, 911).

Coincidences often get our attention. The patterns may not mean anything. Nevertheless, there will be readers who will try to discover the design underlying the seemingly unintentional, arbitrary facts, the "petty details," to echo Kugel, that cry out for interpretation.

Chapter 10

DETERMINISTIC FORTUITY

From where, then, does our feeling of beauty come? From the idea that the work of art is not arbitrary, and from the fact that, although unpredictable, it appears to us to have been directed by some organizing center of large codimension, far from the normal structures of ordinary thought, but still in resonance with the main emotional or genetic structures underlying our conscious thought.

–René Thom

Intervening the gap between analogical determinism and a teleology based on what I call "deterministic fortuity" is a world view known as *mechanical determinism,* which, if it doesn't preclude a notion of divinity, associates God with orderliness and predictability and equates His actions with physical laws, which simply *are* out of necessity, and are not accounted for by the purposes they may serve. In a mechanically determined universe, all matter acts in a physically continuous series, and each event is entirely predictable (in theory at least, although actual prediction may not be practicably possible). As in an analogically determined universe, the future is already given in the present—not due to divine simultaneity in this instance but to linear causality. This causes that, which causes that and so on. Everything is in some sense physically predetermined. The appearance of chance is merely an illusion brought on by lack of knowledge of causal factors. The defining moment for this view came when Pierre-Simon Laplace (1749-1827) asserted that anyone who had knowledge of the forces in nature and position of every thing in the universe at one instant could predict all future behavior. Material determinism obviously raises questions about free will and one's ability to act with purpose.

A teleology based upon *deterministic fortuity* accepts the tenets of material determinism, but also makes use of the concept of *fortuity* in an attempt to argue for human, or even cosmic, agency. Fortuity is nothing like a-causal chance; it involves the coming together of *unrelated* factors that together result in a particular state. For instance, various genetic *and* environmental factors determine one's height. *Telos* is thought to be what actively and continuously causes the various factors to be present in the right proportions so that the state, say five foot two inches, will result. Coincidental factors, the genetic sequences for femur length and a diet containing a certain amount of calcium, would have been *inscribed* in the initial conditions of the universe. The separate causal chains that ultimately converge to produce a specific height would have unfolded according to a finely-tuned plan. This notion of fortuity, unlike the concept of "chance" as I use it, does not explicitly depend upon stochastic resonances involving idea-like qualities of likeness and nearness. Fortuity simply involves the coincidental intersection of distinct causal chains and is more non-mentalistic than mentalistic.

The concept of deterministic fortuity is an old one, dating back to the Enlightenment, but today it is seen through postmodernism and colored by contemporary metaphors. Final causality of the deterministic fortuity variety is sometimes compared to a computer program. Accordingly, separate causal chains have a number of decision forks (1 or 0) as they unroll their ways through time, and although the "decision" at each fork along the way might involve some degree of chance (10% or 30% but never, presumably, 50% or precisely equal, which might lead to forbidden under-determinism, somehow) the odds are actually biased in favor of the design of the original program, which has the inevitable ending encoded in the beginning. Narratives written under this paradigm would *not* interpret coincidental patterns, as in analogically determined narratives. However, predetermined coincidences, intersections of ordinary causal chains, synchronized according to a prespecified design, would be used to further action. A picaresque novel that turns on coincidence and fateful meetings at crossroads follows this scheme, and is critiqued by postmodernists for seeming too contrived, *i.e.* too "teleological."

However, the computer program metaphor does this conception of teleology some injustice. To understand this teleology one must *not* imagine that the initial conditions can unfold a fully specified set of detailed instructions step-by-step in time. One must imagine instead that a dynamic set of constraints and general rules become activated through the connectivity and interactivity of the system they guide. That is, these teleologists thought telic phenomena are self-organizing. (In the 19th century the activated "rules" underlying self-organizing processes were not understood as evolving from stochastic resonances, as I understand them, but were inexplicably already existing.)

The ideas behind deterministic fortuity may be traced back to Immanuel Kant (1724-1804), and were further developed by a number of historically notable figures among whom Ralph Waldo Emerson (1803-1882) stands out for me. English translations of Kant record the first uses of the term "self-organized." He claimed *telos* is given in the interactions between parts and the whole, and final cause is not contained in any thing like an actual physical program that can be separated from the system it guides. According to complexity science philosopher Alicia Juarrero, "Kant's emphasis on recursive causality, wherein the parts are both cause and effect, precludes the existence of a preexisting whole" ("Self-Organization" 113). And as Ernst Cassirer explains, the Kantian whole is "contained in them [the parts] as a guiding principle" (335). Evelyn Fox Keller has also noted that the computer program as metaphor for Kantian *telos* is inappropriate *unless* one has in mind parallel processors, neural networks, distributed networks, and multilayered programs engaged in feedback (see *Century* 103-111). Perhaps as we become more familiar with newly developing kinds of computing we will be better able to imagine the teleology of deterministic fortuity with computer metaphors.

KANT AND EMERSON

Kant was known to have been a well-organized and extremely predictable person, and it is often said that the people of his hometown Königsberg set their clocks to his comings and goings. Whether this is true or not, in his writings he is very disciplined and careful. Emerson was a Unitarian minister, a religious rebel whose writings are wild and passionate, though cerebral. They are both associated with theories of transcendence, which held that the spiritual world was not knowable through empirical study of matter. Instead, the intuition allowed the perception of the spiritual world, and a sense of the *relations* of things, of how objects are structured in space and time. Knowledge that transcended the physical world concerned universals, archetypes, and essential natures.

The reluctant father of German idealism, Kant refuted the radical subjectivity of his follower Johann Gottlieb Fichte (1762-1814), an anti-Semite with strong ideas about nationalism ("*Erklärung*"). Nevertheless, it is Kant's nearness to such philosophies that initially made me uncomfortable with his work. At first I did not like Emerson for similar reasons. This teleology is typical of 19th century thinking, and as a narrative of progress by means of fortuity, it happens to fit well with Social Darwinist interpretations of evolution. Although one cannot say that Kant or Emerson would have endorsed such views, there is a potential association that is difficult to ignore.

I was a sophomore at college when I first read Kant's *Critiques* and Emerson's essays on fate and nature. I was in my Philosophical Naturalism stage and hadn't yet converted to teleology. I couldn't understand how Kantians could come so close to a strictly secular view of the world (empirical proof of God is impossible, God's existence is not self-evident), then swerve to transcendentalism, which left the back door open for a deity. The move (to assert our intuition of metaphysical truths must be *given* if not provable or self-evident) seemed so unexpected and unnecessary. It wasn't until I came to understand this teleology as involving deterministic fortuity that I could understand that it is not a God that might steal into the house, but a form of directional final cause, arguably a very secular form of metaphysical belief. The way in was through fortuity, and that idea, I later realized, was really the only way of hanging on to teleology within a mechanistic paradigm, and it's not altogether wrong. In this teleology god is an abstraction, not moralistic so much as formalistic. It is really simply an argument for self-organization. As Kant writes in an early work (1755), which may highlight some of his driving assumptions,

> *Matter, which organizes itself according to its general laws, produces ... through a blind mechanical process, good consequences, which appear ... design[ed].... When left to themselves, air, water, and heat produce wind and clouds, rain, and streams, which irrigate lands.... However, they produce these results not through mere ... accident (which could have just as readily resulted in disaster). ...[T]hese consequences are limited by natural laws ... to work ... this way. What should we then think of this harmony? How [could] ... things with different natures ... strive to work in cooperation ... unless they recognized a common origin ... in which the essential interrelated construction of everything was planned? If their natures were necessarily isolated and independent, what an astonishing contingency that would be ... how impossible it would be that with their natural efforts they should mesh..., as if an overriding wise selection had united them.* ("Preface")

In Kant's day there was no nonlinear dynamics theory, and so with his teleology he attempted to create one with the inadequate tools he was given. Today we realize that self-organizing trends only seem improbable if you gauge them by statistical models. Kant had only statistical models, measures of randomness, which are useful to those wishing to guess the behavior of linear systems, whose trends are predictable. The more unpredictable organic world seemed to Kant to require a guiding principle to precisely synchronize numerous mechanically determined events. As he writes in 1784,

> *...it may be held that from an Epicurean concourse of causes in action it is to be expected that the States, like little particles of matter, will try by their fortuitous conjunctions all sorts of formations, which will be again destroyed by new collisions, till at last some one constitution will by chance succeed in preserving itself in its proper form,—a lucky accident which will hardly ever come about!* ("Idea" 29)

I associate this form of teleology with what I call *deterministic* fortuity because teleological events are said to be pre-specified at the beginning of time, even if the actual mechanisms of final cause are said to rely on contextual interactions that only unfold within time. It's a very paradoxical theory—self-organizing and yet pre-specified?

From our perspective, Kantian teleology may look too rigid to allow for truly purposeful behavior. But we can attribute this to the classical determinism against which the 19th century teleologists tried very hard to create a new way of imagining teleological activity. They had inherited predeterminism from the mechanistic hypothesis, and the telic part of deterministic *fortuity* is the part that concerns the apparent improbabilities of coincidental intersections.

Teleologists after Kant, who believed in deterministic fortuity, usually did not believe that an intervening God is supposed to exist beyond time, external to the human world. Instead, a nonphysical cosmic intention was believed to be immanent in physical events themselves. The rise of the science of statistics in the Enlightenment had allowed 19th century teleologists to make some theoretical use of chance. Their incorporation of statistics improved upon Aristotle's non-mentalism by including some consideration of the role of chance in developing trends in nature. Consequently, the medium through which a 19th century god-like hand appeared to move was not matter but accumulative coincidence. The role of *telos* was to balance the odds. Although each individual event was mechanistically determined, if one pulled back and took a longer view of things, an orderly sense of nature would appear to reign.

Even though none may be capable of predicting the outcome of particular events—how someone's life turns out or ends, for example—one can estimate the general probability of certain *kinds* of events. We can predict with a fair degree of certainty, for instance, how many pedestrians will be struck by vehicles in a given year. For example, let's say a woman "decides" to take the shorter route home and starts to cross the street, her hat blocking her line of vision. A driver going through the intersection makes a last minute "decision" to turn right and hits the woman. This event involves two distinct mechanistically determined causal chains, but to 19th century teleologists, it is also providential, statistically speaking. Although an individual death is unpredictable, because the overall number of deaths in a given year is statistically fairly certain, the individual death would seem to be predetermined. Teleologists invested in the concept of deterministic fortuity imagined the initial configuration of "particles" (atoms) had been precisely chosen so that the timing of coincidental events in the distant future would be perfectly synchronized causing the apparently "chance" effects. Although Kantian teleologists did not suppose that God worked like a puppeteer, bringing specific pedestrians and specific speeding vehicles together on wet roads, they did believe there was an internal principle that spontaneously guided natural processes so that, on the whole, there emerged a kind of probabilistic order that was inevitable.

In the 1860 essay, "Fate," writing about "the new Science of Statistics," contemporary events and celebrities, Ralph Waldo Emerson observes,

> *It is a rule that the most casual and extraordinary events—if the basis of population is broad enough—become matter [sic] of fixed calculation. It would not be safe to say when a captain like Bonaparte, a singer like Jenny Lind, or a navigator like Bowditch would be born in Boston; but, on a population of twenty or two hundred millions, something like accuracy may be had....* (950)

Emerson goes on to say that these are:

> *hints of the terms by which our life is walled up, and which show a kind of mechanical exactness, as of a loom or mill in what we call casual or fortuitous events.* (951)

Although Emerson takes some comfort in the belief that order—arising out of so many fortuitous events—enabled the intuition of orderly purposefulness, he also seeks to argue against the loss of human freedom that this "exactness" implies. While accepting the fact that mechanistic determinism results in "form," "fate," and "limitation," he also argues that there is a way to conceive of activity that results in "power," original "thought" or agency, and "freedom." According to Emerson, an individual particle of matter is constrained and limited by mechanistic laws, but these laws only apply to individual elements *in isolation*. The synchronization of events in their total relation is *not* determined by mechanistic laws. Likewise, telic forces do not guide the *individual* life per se; instead they guide the entire collocation of matter according to an unknown and unknowable rational plan. The individual has a kind of divinely bestowed power or agency insomuch as he participates in that plan and thereby transcends material causality. But the ultimate transcendental culmination is not intended by the individual; intentionality is conferred upon the individual by means of grace. One might say it is a by-product of cosmic *telos*. Emerson's notion of "power," the complementary aspect of mechanistic "form," is a product of chance, not choice:

> *Power keeps quite another road than the turnpikes of choice and will: namely the subterranean and invisible tunnels and channels of life.... Life is a series of surprises...* (482)

Although the 19th century teleologists were, in some sense, mechanists, they were not reductionists. They also believed there was something more. They believed that, to use a favorite phrase, *the whole is more than the mere sum of its parts*, which gets at the difference (they didn't fully comprehend) between predictable statistical trends and unpredictable self-organizing trends. Emerson believed that it was useless to try to infer the overall organizational strategy from mechanistic laws. In his 1844 essay, "Experience" he writes,

> *How easily, if fate would suffer it, we might ... adjust ourselves, once for all, to the perfect calculation of the kingdom of known cause and effect.... God delights to... hide from us the past and the future.* (482)

Because this is so, the "results of life are uncalculated and uncalculable" (483). Thus, Emerson mocks whatever reductive science happened to be the fashion of the day:

> *I hear the chuckle of the phrenologists...who [know the law of a man's] being; and by such cheap signboards as the color of his beard or the slope of his occiput, [read] the inventory of his fortunes and characters.* (475)

Emerson recognized that the ability of predicting statistical trends does not pertain to individual purposeful acts, which are complex and emergent, though he was not quite able to put it this way.

Caught up in this teleology, with its inherent defects and difficulties, are such thinkers as Herbert Spencer (1820-1903) and Karl Marx (1818-1883) who both thought history was progressing toward a certain goal. I will say little about their theories, but I cannot help but stop to note how very different political ideologies can come from deterministic fortuity. On the one hand, you have the social biologists who would try to help progress along a bit, assuming wrongly that one can use statistics to predict and control complex events. On the other hand, you have those who believe life is far too complex to be tinkered with successfully, and instead one ought to allow anarchy to reign with the mistaken belief that things will self-organize in the best possible way. Emerson's view on this issue is closer to my own: a sensitive artistic temperament is the best guide for political action.

But for all his talk of freedom and power, Emerson is still a determinist, an unavoidable outcome of the context in which he and other 19th century teleologists wrote. In the teleology of deterministic fortuity, human purpose is, in some sense, only apparent. Whatever original thought a human being may have, and whatever modicum of agency this would seem to supply, is actually always already prescribed in the divine plan. Thus, "fate slides into freedom and freedom into fate" (961).

In "The Dilemma of Determinism" (1897), William James, Emerson's nearest intellectual descendant, attempts to integrate the concepts of chance and freedom in a determined universe. James associates chance with freedom of action and subjectivity. In his *pluralistic* universe, synchronization is not necessarily so exact. He argued that there are a number of ways to arrive at the same conclusion. Another way to put it, emphasizing his indebtedness to Darwin, is to say that there existed "chance variations" within the realm of possible action, and the fittest of these are selected again and again in time. James illustrates his theory by imagining life as a game of chess between the novice man and the expert God:

> *Suppose two men before a chessboard,—the one a novice, the other an expert player of the game. The expert intends to beat. But he cannot foresee exactly what any one actual move of his adversary may be. He knows, however, all the* possible *moves of the latter; and he knows in advance how to meet each of them by a move of his own which leads in the direction of victory. And the victory infallibly arrives, after no matter how devious a course, in the one predestined form of check-mate to the novice's king.* (181)

In James' view then, there are a number of possible fortuitous relationships that might result in the same end. This, by the way, is consistent with what we now know to be true in biology in the case of many-to-one genotype to phenotype mappings and in nonlinear dynamics in the case of the limited number of structural archetypes that arise out of large number initial configurations. James is gesturing toward the idea of directionality; he is not describing true purposefulness that can only come with an admixture of originality.

DYNAMICAL STABILITY VERSUS PRESPECIFIED FORM

James may be considered a pivotal thinker between deterministic fortuity and pragmatic teleology, which comes in the next chapter. In some sense deterministic fortuity as a concept is a wrong way of thinking, and yet it functions as a step in the right direction. It attempts to naturalize teleological processes. It initially paves the way for and finally incorporates the concept of evolution by natural selection (think of Kant's "wise selection"). It recognizes and tries to explain trends and progress in nature. It has an (albeit incomplete) understanding of the importance of holistic thinking. And although it lacks a precise understanding of self-organizing processes, it offers the first accurate definitions of self-organization.

The contributions that this teleology makes to science were largely rejected during the 20th century due both to prejudices against teleology in general and the specific faults of this teleology in particular. The defects of this teleology, however, are more due to the determinism that reigned during the late 18th and 19th centuries, and continued to some degree in the 20th century.

I want to note a great and interesting irony that we can perhaps understand the defects of this teleology better by comparing them with the defects of 20th century genomics, which were plagued by comparisons of DNA with fully-specified blueprints unfolding in time. The worst fault of the teleology of deterministic fortuity is that *telos,* like DNA, is too easily misconceived as a pre-existing material code (like cosmic DNA in the initial configuration of the universe) that determines all future events and actions.

During the past century, as Evelyn Fox Keller has noted in *Century of the Gene*, the cause of biological form was thought to actually reside in the physical structure of DNA, and similarly, some 19th century teleologists had supposed, counter to Kant, that the future was physically inscribed in the initial conditions of the universe. According to reductionist genomics, traits and characters would be in some sense materially predetermined. In the 21st century, genetic research is turning to a more holistic understanding of organic development.

While the Human Genome Project had promised to reveal the blueprint that determines what and who we are, now that the project is complete, researchers in the field have found that the gene does not, in and of itself, contain an explicit and detailed code for all of development. The secret of life is not to be found in the molecular basis of genetic information, and it is not a simple matter of "decoding" the message of DNA. It turns out that the physical structure of a gene does not necessarily give an exact indication of what that structure eventually will lead to. The functions of genes *and* of the proteins produced through the process of genetic transcription depend on context. The transcription process itself is largely determined by the chemical environment of the cell in which it occurs. Consequently, there is no simple relationship between gene structure and protein, and there is no simple direct linear relation between any given protein structure and its function. The stability of traits through generations, as Keller has noted, is no longer attributed to a gene as "an inherently stable, potentially immortal, unit that ... [is] transferred intact through generations" (14). The *stability* we once attributed to the physical object known as a "gene" has now been found to be a product of the *dynamics* of development, which involve the evolutionary acquired meanings/functions of types of things to types of organisms. The purely reductive mechanistic view of genetics is dying. In its place is developing a holistic approach, which is referred to as "functional genomics" rather than "structural genomics." According to Keller, in light of recent findings, Kant's teleology seems a better description of the process of organic development than does the reductive view held by genetic determinists (106-110).

Hindsight helps us see Kantian and Emersonian teleology anew. The techniques developed late in the 20th century for understanding self-organizing process removes the biggest stumbling block to this teleology, the inbuilt predeterminsim around which it tried so hard to maneuver.

I have already described, in the sections on directionality, the scientists who have revived some of the ideas behind this teleology, involving formalisms, directionality or archetypes. In literature such revivals were not quite as necessary, very generally speaking, because holism, interconnectivity, fortuity, and formal organization are inherently part of the artistic process. In the next section, I show how Czech novelist Milan Kundera makes the concept of deterministic fortuity explicit in his fiction. Playing with 20th century computer metaphors and the tenets of genetic determinism, which themselves are understood with computer metaphors, Kundera makes, consciously or not, a Kantian critique of 20th century reductionism.

KUNDERA AND DETERMINISTIC FORTUITY

In his 1990 novel, *Immortality*, Milan Kundera writes, "All power of decision has been left to chance" (12). Like the transcendental teleologists, Kundera brings in the notion of fortuity in order to escape the reductiveness of classical determinism. In accordance with the mechanistic hypothesis, Kundera also attributes unpredictability to human ignorance; behind this *apparent* unpredictability there would exist a calculation so complex as to be beyond our understanding, but to a super scientist, who had knowledge of the initial configuration, it would be wholly determined and predictable.

When considering Kundera's notion of teleology and causality, it is a telling fact that a favorite book of his is Denis Diderot's *Jacques the Fatalist* (see "An Introduction"), written 1771-1773 in the nursery of material determinism. Kundera also wrote a play, *Jacques and his Master,* in 1985. Although Diderot's Jacques is committed to a belief that all is foreordained, he realizes that life must be lived in ignorance of what the inevitable outcome will be. When asked why he decides to act as he does in any given situation, Jacques replies:

> *Why? My God, I don't know.... Without knowing what is written above, none of us knows what we want or what we are doing, and we follow our whims*

which we call reason, or our reason which is often nothing but a dangerous whim which sometimes turns out well, sometimes badly. What man is capable of correctly assessing the circumstances in which he finds himself? The calculation which we make in our heads and the one recorded on the register up above are two very different calculations. How many wisely conceived projects have failed and will fail in the future! How many insane projects have succeeded and will succeed! (28-29)

In Kundera's novel *Immortality*, we have a character who might be called "Agnes the fatalist," and final cause is represented, not as a "calculation," but similarly as a "program" in "the Creator's computer," and life as,

a play of permutations and combinations within [this] general program, which is not prophetic anticipation of the future but merely sets the limits of possibilities within which all power of decision has been left to chance. (11-12)

Unlike the Medieval universe determined by analogies, in Kundera's materially and fortuitously determined universe, one would not have to be a "poetic" prophet in order to predict the future: one would merely have to know the initial conditions, which would give the odds. Though each decision might appear to involve some degree of chance, the cards are actually stacked in favor of the design of the original "program." Each separate causal chain unfolds according to a finely-tuned plan; the coincidence occurs on schedule and has its intended effect. Although Kundera uses the metaphor of the computer program, his teleology, consistent with deterministic fortuity and possibly new nonlinear forms of computing, seems more like a dynamic set of constraints and general rules that become activated through the connectivity and interactivity of the system.

Immortality is a metafictional story of Agnes, who is married to Paul, and her sister Laura, who is involved with Paul. It's really about a character called "Kundera," who is writing a novel called *Immortality*. Exploring the common human desire for immortality through fame or lasting effects, the novel digresses to relate Goethe's personal history and conversations (in the afterlife) with Ernest Hemingway. There is also a Professor Avenarius who discusses story writing with Kundera. At the end of the novel, Agnes attempts suicide (her married life is without charm) and sits down on a busy roadway. She isn't killed, but she causes an accident.

Not terribly plot-driven, the novel is largely philosophical and includes many reflections about a teleology that would be based on deterministic fortuity. In Kundera's fictionalization, we find ideas reminiscent of 19th century Kantian teleomechanists, who held that there were archetypes from which individual organisms were derived. Writes Kundera,

> *The computer did not plan an Agnes or a Paul, but only a prototype known as a human being, giving rise to a large number of specimens that are based on the original model and haven't any individual essence.* (11-12)

As with a transcendental teleology, telos here does not guide the individual per se, only the overall organization of the whole by means of chance, which here means coincidence.

Immortality includes a long discourse on the nature of coincidence. The discussion takes place between Avenarius and the textual characterization of the author himself, to whom I will refer as "Kundera." He and Avenarius try to categorize various kinds of coincidences and synchronizations for a book that "Kundera" dreams of writing called *The Theory of Chance*. "Kundera" has an aesthetic prejudice against novels that follow one causal chain without showing how that chain is inexorably entangled with many others and the whole. "Kundera" criticizes the novel that rushes "toward a goal" in the way that he says Aristotle recommends (for playwrights).

> *I regret that almost all novels ever written are much too obedient to the rules of unity of action. What I mean to say is that at their core is one single chain of causally related acts and events. These novels are like a narrow street along which someone drives his characters like a whip. Dramatic tension is the real curse of the novel, because it transforms everything, even the most beautiful pages, even the most surprising scenes and observations merely into steps leading to the final resolution, in which the meaning of everything that preceded is concentrated.* (238)

In an act of rebellion against Aristotle's unity of action, "Kundera" announces that in part six of *Immortality* he will introduce a character who "causes nothing and leaves no effects" (238). Indeed part six at first appears to be a digression from the main action, describing how a new character named Rubens meets a young girl who dances in such a way that he decides to call her the "lute player." In the course of this digressive tale, it is described how he loses track of her, but thinks about her often. Years later, they meet again by chance and then become occasional lovers for many years.

But this gratuitous figure, Rubens, proves to be significant to the main action after all. In a later chapter, we learn that the lute player is Agnes. Rubens' story actually has a function: it provides information that contributed to the heroine's state of mind, which ultimately leads her to wish for death. I don't see my friend Aristotle minding Rubens' inclusion much. While the ancient Greeks were inclined to see one's nature as fully given, since around the time of *Hamlet*, the west has been more receptive to the notion of psychological development, and a person's various experiences are relevant to the main action insofar as they contribute, even indirectly, to that person's developing character. The Ruben's story may add another causal chain to *Immortality*—a chain that is more psychological than material—but it is not irrelevant. The story of Rubens is significant *because* it appears in a narrative determined by deterministic fortuity. Everything counts: everything in life is related to the whole. Life, "Kundera" says,

> *does not resemble a picaresque in which from one chapter to the next the hero is continually being surprised by new events that have no common denominator. It resembles a composition that musicians call a theme with variations,*

and, he adds, "you won't escape your life's theme" (275). The way "events are synchronized" insures this (225). Our lives may seem picaresque: disconnected, merely random or episodic (300), but, these random conjunctions turn out to be fateful insofar as all is connected to a whole. Says "Kundera,"

> *If our lives were endless... the concept of episode [essentially unrelated series of events] would lose its meaning, for in infinity every event, no matter how trivial, would meet up with its consequence and unfold into a story.* (305)

Milan Kundera may have written in a time of postmodernism, but his narrative aesthetics conform to the sensibilities endorsed by 19th century teleologists who in some sense extended and revised Aristotle's non-mentalism. The narrative of deterministic fortuity is more complicated than the Aristotelian narrative—there are many more causal chains—but it is definitely a descendant, for even though chains may interrelate fortuitously they themselves are each directly and materially connected. There aren't any gaps in these chains, and there aren't any wholly new chains. There aren't any miracles. "Kundera" is definitely more Aristotelian than Augustinian.

CONCLUSION

Teleology of the 19th century was deterministic because it was built on the foundations of material determinism, but *telos* itself was found in the dynamic interactions of a system and emerged in time. Because these teleologists were primarily interested in the universal laws that guided holistic systems in terms of the functional relationships of the parts to the whole, the kind of telic phenomena they studied tended to have the aspect of what I have referred to as *directionality*. So long as the strong mechanistic hypothesis held, there was no real way to make an argument for the aspect of telic *originality*. The conclusion is unavoidable under classic determinism that everything is always already extant. Nothing is wholly new; rather all things are different permutations of old arrangements or variations on a theme.

Chapter 11

PRAGMATISM

We must therefore suppose an element of absolute chance, sporting, spontaneity, originality, freedom, in nature.

–Charles Sanders Peirce

In the twentieth century, a teleology informed by pragmatism became possible. Although this reinterpretation of teleology owes much to William James, he still worked within the framework of deterministic-fortuity teleology described in the last chapter. According to that view, when one acts subjectively, he or she participates in a fortuitous arrangement that had been predetermined by universal laws at the beginning of time. Humans, in this case, are not really what I would call "free," but their subjective acts help create a harmony that had been determined by a particular arrangement of matter at the beginning of time. In order to incorporate a notion of true originality in teleology, we need to think of matter as essentially indefinite and determined by contextualized interactions. For these reasons, the pragmatism discussed here belongs more to James' friend Charles Peirce, whose notion of pragmatism (or "pragmaticism" as he called it) entails a belief that the initial state of the cosmos was indeterminate and that the interpretation of signs, which have different meanings in different contexts, have real physical consequences in ways not accounted for by classical material determinism.

The twentieth century began auspiciously with the acceptance of Planck's constant in 1900 and the subsequent introduction of quantum mechanics in the late 1920s. Planck's discovery, as Niels Bohr writes, imposed "upon individual atomic processes an element of discontinuity quite foreign to the fundamental principles of classical physics" (4). With the introduction of quantum mechanics and the indeterminacy of sub-atomic particles, the 19th century notion of *telos* as consisting in the order and arrangement of the original configuration of well-defined particulate matter was no longer tenable. Thus, *telic* activity could not be exactly prespecified. Far from making teleology impossible, as my friend Arkady Plotnitsky argues (see Chapter Three), this development removes the difficulty 19th century teleologists had tried so hard to overcome. This is not to say that quantum mechanics explains teleology or that quantum indeterminacy causes purpose. Not at all. These discoveries in science simply removed the mistaken requirement imposed upon 18th and 19th century teleologists that *telos* need be strictly prespecified—since, apparently, *nothing* is. It also reinforces the idea that purpose emerges in the individual self-organizing entity and is not imposed upon that entity by any pre-existing outside agency. A puppeteering god with a fully specified plan has no role to play in the 20th century. Intentionality, according to a teleology that incorporates pragmatism, is not innate in the thinker, as some 19th century thinkers would have had it. Rather, the ability to think freely (to think about things possibly non-existent, for example) is an evolved "gift," as it were, compliments of the thinker's dynamic situation.

Throughout most of the 20th century, however, pragmatic teleology remained largely just a possibility. Early on there was new talk of "emergence" in science—especially chemistry and system science—that had not been broached before (See Goldstein, "Emergence as a Construct"). In philosophy there was the influential emergent evolution of Henri Bergson (1859–1941) and the process theology of Alfred North Whitehead (1861–1947). In biology there were the controversial non-Darwinian theories of Richard Goldschmidt (1878–1958) and C. H. Waddington (1905–1975). In the 1960s, James Lovelock introduced the concept that our environment is self-directed like an organism. Autopoietic theory made fledging flights. However, many of these new ways of thinking tried to *avoid* an association with the name "teleology" with its 19th century taint.

Meanwhile, classical science grandly withstood the quantum assault on determinism, since above the quantum level all simple physical systems behave as classical physicists always said they did, and even the parts of complex physical systems do too, as long as they are taken out of context. So then at the very moment pragmatic teleology becomes possible, it actually got shoved rudely aside. Notwithstanding the great contributions of Bergson and Whitehead, in the 20th century, more than in any previous century, teleology becomes the object of suspicion and derision.

Nevertheless, I say teleology was sorely *missed*. Its exile was not ordered without a great deal of misgivings and regret, from artists especially, who knew, who *knew* there was something to it, despite the fact that science was telling them otherwise. I propose that a teleology based on pragmatism characterizes the period of High Modernism in literature and explains the strong sense of a yearning that is often sensed by readers. Yeats's falcon cannot hear the falconer. This sense of loss, this largely unfulfilled desire, has been misunderstood by some as the Modernist's nostalgia for the essentialism that was no longer defensible. But theirs was no grief over mathematical and metaphysical axioms that lay shattered. I say it is teleology they missed, not essentialism. This is not to say that the argument of pragmatic teleology is explicitly there or that these artists were bent on inculcating a clearly defined metaphysics; it is *felt*, however, in Wallace Stevens, late Henry James, Virginia Woolf, and Vladimir Nabokov, to name the few that taught me pragmatic teleology. A hallmark of this type of narrative is the idea that selfhood, while not essentialist, is still possible. In High Modernism there is a strong sense of formal order that is emergent and purely aesthetic, and this has its roots in deterministic fortuity, which the pragmatists graciously inherited. Additionally, there is playfulness, a love of coincidence and an appreciation of the kinds of gifts that mistakes can bring. As James Joyce writes, "A man of genius makes no mistakes; his errors are volitional and are the portals of discovery" (*Ulysses* 9.228-9). Especially notable in Joyce, Stevens, and Nabokov, among other Modernists, chance highlights the important role of interpretive contexts in pragmatic teleology.

In these next sections, I will present a definition of radical novelty that is made possible by the complexity sciences. It will help explain, I think, what the High Modernists were doing. This is not to say that High Modernism is the end point of literature or that contemporary writers ought to return to that model. The Modernists no more had the complete picture than Aristotle or Augustine or Kant. There is no complete picture, and the picture is always changing. In the last chapter, I will present what I think are some of the ways that teleology may be developed anew. And who knows what may come after that. But I do believe that Modernist literature, insofar as it evinces pragmatic teleology, incorporating aspects of Aristotelian, Augustinian and Kantian thought, is a development, is progress.

RADICAL NOVELTY

What is radical novelty? We can only start asking such a question if we do not believe in predetermination. I've already begun the discussion in Chapter Two with my ╥ and ╙ example in which I showed how order *that is not prespecified* can emerge. This emergence is new, but in that chapter I was more focused on self-creation, maintenance and organization, or directionality. Here I want to focus on *changes* in directionality, that is, originality.

I will turn again to some very abstract models, not unlike the ╥ and ╙ one, in order to define the notion of *structural complexity,* which needs to be done before we can proceed, since I am going to use it to define "newness." In the complexity sciences, when an entity emerges from a self-organizing process, we say that its structural complexity is what makes it different from the parts that went into making it. My simple model will be used to describe patterns that exist in various kinds of systems, physical, biological, social, and aesthetic, and that's why it will have to be as abstract as possible. With such a form of representation, we can consider the patterns, forgetting about the content.

For comparison, let's look at a couple of simple processes first, not complex ones. 1.) A fair coin randomly comes up either heads or tails. 2.) A heating system is activated/deactivated when its thermostat senses that room temperature goes below/above fifty degrees. We can describe the fair coin in terms of a binary code of 1s and 0s in a random sequence: 0100100010100101.... We can describe the states of the heating system as 10101010101010.... You might think the fair coin is more complex, but it is only random. Neither of these processes is structurally complex. If I were to try to guess the next outcome at any point in this sequence, my ability to predict would never improve the further I looked along the sequence: in the coin toss sequence, there is always a 50 percent chance of correct prediction. In the heating system, there is always a 100 percent chance the next element will be different, since it oscillates back and forth, and so prediction is always 100 percent correct. Structurally complex process are different from these two simple processes in that some one who is studying the behavior of a complex system will be able to improve his/her guesses about what the next state (of a part, if not the whole) will be the more the person observes the system.

We can also try to describe the behavior of a complex system using a binary code. Let's say for example we have a complex system—it doesn't matter what the system is, a cell changing states, a child inventing a simple tune—that can be described with the following string: 11101011101011101 0. We notice that, for instance, 0s are always followed by a 1, or that either a single 1 comes between two 0s, or three 1s occur together. Based on this understanding of the pattern, if we begin observing somewhere in the middle of the process and we see a 0, we can predict that a 1 will be next. Based on the same understanding of the pattern, however, if we see a 1, we are not so sure what will come next. There is equal probability that we will observe either a 1 or a 0 next. If then we see a second 1, however, we can predict with more certainty that a 1 will follow.

1s are "different" because the complex process has some historical memory. In just this way, structurally complex processes mix occasional predictability with occasional unpredictability and require more elaborate descriptions than completely unpredictable or completely predictable processes. A digit in a structurally complex string is a bit like a letter in a specific language: if one sees a Q in English one knows a U will follow; if one sees an S, one knows that an X or Q will not follow. While additional exposure to a simple system will not increase our ability to predict its next state, with a complex system, we can become more certain the more information we gain about a complex processes. The complex process involves the concept of memory and determines its next state based on the immediately proceeding state. This notion of structural complexity has been developed by Jim Crutchfield as a means of *quantifying* and *categorizing* self-organizing processes (see "Calculi of Emergence").

The observer of a complex system learns about it by interacting with it and becoming changed by these interactions. The observer becomes part of the system, in a sense, and therefore can act as a (partial) model of it. The observer of a complex system cannot become an accurate predictor in the usual sense, but he can become a very good guesser. The observer (or interactor) interprets the indices with which he (or it) interacts and also becomes an index.

That's all well and good, but we haven't yet started to discuss radical novelty, which I am still not quite ready to do. One more step first before we can begin.

Now if we take our structurally complex sequence 1110101110101110-10111010 and replace some of the 1s and 0s with *s that indicate *wildcard* positions, then we have more plasticity. A wildcard can differ from the prescribed 1 or 0 without affecting the overall outcome or the meaning, we might say, of the process. These systems with wildcards are less rigid than the example above. Now we have a musician who respects the notes as prescribed, but can play them at different octaves or substitute chords for single notes. We can describe such a process this way: 111*101*1010*1101*111*10. Even though the wildcard positions (where the musician varies the performance within a limited range) seem surprising, we can still understand the essential properties of the process. We recognize the tune. This is the kind of plastic behavior or *pluralism* that William James imagined was possible in a world that was teleological in the sense that it followed a fairly predictable pattern (having the aspect of directionality), but was also open to an agent's subjective interpretations of what might define a 1 or a 0. To give some examples of plasticity and pluralism from biology, different genetic sequences can produce essentially the same wing pattern on different species of butterflies, for example, the monarch and the viceroy (different means produce the same end); or, conversely, a variety of races with different wing patterns might be considered the same species of butterfly (differences are ignored and generalizations made). Allowing for the existence of wildcards means that even if there are departures from the norm, there can still be order. The freedom of wildcards is meaningless and may be overlooked. This is partly why we experience what I have referred to as telic *directionality* in previous chapters. The question we want to look at in this chapter, however, is, *What if noise did matter?* Now we get to the subject of radical newness.

In artistic creations, noise matters. One might say this is the essence of art. I will try to make the case, using Henry James later in this chapter, that teleological order—since it can't be predetermined and it isn't strictly materially determined, and it's not a-causal—must emerge out of something like error, chance, coincidence. It must involve interpretation. So rather than subjectivity having only a benign effect on teleological design (not disturbing it too much), it has the utmost importance, and can help create something entirely novel.

MEANING AND NOVELTY

Now that we have a way of formally comparing structural complexity, we may return to the question of how "newness" emerges. Organisms that operate with structurally complex models of the world that include wildcards or some degree of noise have more potential to adapt than organisms whose models of the world are simple, rigid and clearly defined. Their subjective acts matter and can change meanings of structures/patterns in themselves or in their worlds. The fittest organisms are able, not just to act in a regular way, making generalizations despite noise, but are able to be mistaken in their generalizations and thereby make creative use of noise. The role of error—of misinterpretation, of presuming what was not the usual meaning—is essential to our present-day conceptions of human intention and creativity (see Boden).

If we consider any individual thing out of its usual context, it becomes apparent that it has no inherent meaning or function. This is true even in biology, where, presumably, things exist in certain ways because of the survival function they have served, leading to increased reproduction. In this way, natural selection has created a "natural" or "correct" meaning for a structure—a longer beak or a cell's particular protein receptor—but that meaning is only "correct" from the perspective of the organism in which it evolved. As Ernest Nagel argued in the 1960s, it is misleading to say, for example, that the *only* function,

> *of the white cells in the human blood is to defend the human body against foreign microorganisms. This is undoubtedly a function of the leukocytes; and this particular activity may even be said to be the function of these cells from the perspective of the human body. But leukocytes are elements in other systems as well ... of the system composed of some virus colony together with these white cells...* (422)

From Nagel's remarks we see that one system, a virus, can make a very different use of another system, a white blood cell, than the one for which it developed. A viral appropriation of a pre-existing structure vividly describes a way in which originality can arise. The virus makes the while cells function for it by making their activities mean something different. I consider this a creative act insofar as it involves error or a distortion of the rules, and it requires an overcoming of the former ways of understanding facts.

And to turn the coin over, how much of human DNA came to us laterally through the viruses that infected our ancestors over millions of years? How has the human body co-opted viral genes for its own functions, evolving radically new ways of being? These are questions 21st century geneticists are starting to ask and the answers will likely have to involve a theory of interpretation and biosemiotics.

Another way to examine the notion of creative error in pragmatic teleology is to look at it in terms of the advantageous use of *side effects*. Stephen Jay Gould and Richard Lewontin's "spandrel" is a term used to describe a physical structure that is a side effect of other functional structures that have arisen through adaptive evolution. These spandrels, "adaptively neutral" or meaningless structures, don't themselves serve a function defined by natural selection, for any one or thing, but may at some point become useful to an organism through an evolutionary process known as exaptation. Let's say that the way an animal's hair grows (shaggy, short, or thin) is partly determined by genes, which are selected for some advantage. The same gene may also affect claw, nail, and/or horn development, and so an adaptive change in a species' fur may result in a neutral or non-adaptive change in, say, antler size. In some cases, the new antler or horn may prove to be useful too. These kinds of changes are known as "correlated variations," and they were a focus of 19th transcendental morphologists. When Darwin found patterns he could not explain simply in terms of fitness selection, even he called them "whimsical" correlations—as if some Joker were responsible (*Origins* 5). Although side effects are determined by physical laws and constraints, the creative interpretation of a side effect, such that it comes to serve a function for the interpreter, constitutes original teleological behavior in the pragmatic sense.

The term "side effect" is itself interesting in terms of teleology because when we use the term we usually have in mind an effect that is not intentional. In *Teleological Language in the Life Sciences,* Lowell Nissen reminds us that a side effect is still very much part of determinism. Side effects are not the product of any thing like objective chance, however much ordinary language likes to categorize side effects as "chance" effects. He writes,

> *We say that relieving pain is a function of aspirin but describe stomach irritation as a side effect, and we contrast the function of a machine with its by-products. Terms such as "accidental," "fortuitous," "just happens," "by some*

quirk," "side effect," and "by-product" in such contexts do not refer to what is random or uncaused, but rather to what is not planned, not intentional. (222)

As Nissen makes clear, a side effect is defined *as such* by the particular frame through which it is viewed. If we believe a system's function is to produce C, and an effect of the system does not contribute to C, we call that effect a side effect. But some side effects do become useful. They are like ready-mades just waiting for the right artist.

EXTRINSIC EMERGENCE

As discussed in the previous chapter, the focus of 19th century non-mental teleology was *intrinsic* emergence. There was no external supernatural agent who constantly intervened in order to limit, direct, and guide natural processes: natural processes, it was argued, were self-organizing and self-directing. In this chapter, I have begun to discuss how interaction between one system and another separately developed system can result in telic originality. I have also mentioned how the creative use of a side effect within a system can also result in telic originality. We can refer to this kind of process as exhibiting *extrinsic* emergence. While intrinsic emergences occurs when parts interact to form a whole, extrinsic emergence occurs when two different wholes interact.

To give a more detailed illustration of extrinsic emergence—to capture it in the act as it were—let's turn to another abstract model. I borrow this one from Nabokov. He liked to play a game called "word golf," in which a player starts with a word and changes one letter per move until he has made an entirely new word; each move in the game must produce an actual word: no nonsense words allowed. A "lass" can be turned into a "male" in four moves: lass, mass, mars, mare, male. I use word golf as a basis for my model because it illustrates ruled-bound change. Changes in organic nature, likewise, are also rule-bound and do not permit non-viable organisms. Changes in inanimate physical systems are also rule-bound and do not permit structures that, for example, exceed certain energy requirements.

Let's imagine there are four distinct communities of individuals that each reproduce/interact only with other members of their community. Each community is designated by a word, which represents a string of code. This model is a more complicated version of the binary string model as it has twenty-six rather than two possibilities. It is also more structurally complex as it is constrained by the rules of English spelling and vocabulary.

The four communities are: smiles, catsup, snooze, and bolder. These names will be used abstractly, that is, we are not to think of smiles, for example, as having anything to do with smiling. We merely want to use a code that is constrained by some set of rules. Thus, we borrow English words for their structural properties not for their meanings.

Although we know these communities by these names, each of the individuals in any community is actually coded for by only a few of the letters in its name. For example, the only letters that are important in defining a smiles as such are the third, forth, and fifth, letters, i, l, and e, respectively. The first two letters, s and m, and the last, s, are wildcards. Wildcards are incidental to what is considered the essential nature of the individual, or more neutrally, what that individual has in common with other members of its group or what natural selection has "seen," evolutionarily speaking.

If, for instance, a housing community defines itself by its Tudor style of architecture, the i, l, and e determine the structure of the buildings, while the wildcards determine, say, the color of brick, which can vary. Another community might define itself by surface color. In which case, all red houses, say, would be considered part of the community regardless of the style of architecture.

If we indicate the defining letters of each group with bold face font, we get sm**ile**s, **ca**t**s**u**p**, **sn**ooze, and **bo**lde**r**.

Random mutations in the incidental wildcard letters over time have added diversity to the sm**ile**s population. It is now composed of five individuals: smiles, toiler, stiles, smiles, and smiles. The rules are preserved: they each have ile in the right places; they are each English words. Within the sm**ile**s community, no one is overly concerned with this diversity. One might think of the sm**ile**s community as an ideology, like Christianity, that tolerates (ideally speaking) a variety of sects—Protestant, Catholic, Greek Orthodox, Russian Orthodox—and still retains a basic identity that is shared among the different groups. One might instead think of sm**ile**s as a community of literary interpreters. All individuals agree that, say, *Hamlet* is about doubt, but they each have a different focus, which may include an interest in metaphysical, political, social, familial, or sexual uncertainties. One might instead think of a particular species, like *Homo sapiens*, which includes a wide diversity of peoples.

Diversity within well-defined communities is the norm rather than the exception. However, there are those communities that will not adapt to change. For example, it might be the case that no mutation in the **cat**su**p** community has ever managed to establish itself. **cot**g**tp**, for example, would not be a viable individual according to the rules of English words. The **cat**su**p** community is not diverse: it is composed of three individuals: catsup, catsup, and catsup. The **sn**ooze community has managed to incorporate some diversity: snooze, snooze, snooze, snores, snares, and snooze. The **bo**ld**er** community has accepted one mutation. It is now made up of bolder, bolder, bolder, and boiler.

If we replace each wildcard letter with an *, we can see how members of the same group see each other. When smiles looks at either stiles or toiler, it sees **ile*. The wildcard option allows for hidden diversity in the sm**ile**s community. The members are at liberty to experiment with a variety of letters in the wildcard slots. Some catch on. Others do not. But, in general, wildcard options represent a safe arena for play and freedom. What is probably more important, however, is the way that wildcard options also result in a great deal of conformity. At the end of the day, the wildcard differences are irrelevant to the identity of a community as a species that is separate and distinct from other species. This is directionality and may be compared to Motoo Kimura's 1968 theory of neutral evolution, which concerns many-to-one genotype to phenotype mappings.

The reproductive rule, "members of a community can only reproduce with their own members," prevents interbreeding between members of different communities. However, one day, smiles and boiler—from the sm**ile**s and **bo**ld**er** communities, respectively—meet by chance at the crossroads and interbreed, producing a hybrid mutt soiler. (Okay, so the meaning of these words in English does have a function, to make you smile.)

To understand how the reproductive proscription can be transcended, we need to consider the effects that interpretation can have. Members tend to recognize only what they already know. Everything else is noise to them. They do not pay attention to difference. They have no rule by which they can make sense of noise; therefore it goes unnoticed. For example, when a sm**iles** looks at a **catsup**, it sees ******, since the third fourth and fifth letters do not match the essential characteristic that a sm**ile**s recognizes in a community member. When a sm**ile**s looks at a **sn**ooze, it sees ******. When a sm**iles** looks at a boiler, it, however, sees **ile*.

By the above it is clear that, as far as smiles is concerned, boiler is also one of its own community. It is a coincidence that boiler has an ile wildcard configuration that happens to be interpretable by a sm**ile**s. Although the similarity is meaningful to smiles, smiles is imposing an analogy upon boiler's noise. No single causal mechanism produced both occurrences of iles. In boiler's case ile is just a chance mutation, a singular event, which occurred within the **bo**lder community. Whereas, smiles' letters ile are genetically fixed by natural selection.

I remind the reader that this model is just a metaphor. If such a chance similarity in patterns were to actually happen in the biological world, it would be a rather improbable instance of what is known as *evolutionary convergence*. As Nabokov, who was a lepidopterist as well as a novelist, has explained, convergence is a similarity,

> *attained by essentially different means. Such false resemblances are extremely rare and the number of characters involved is small, and this is as it should be, since such "convergence" depends upon the mathematics of chance.* (Qtd in Boyd *Nabokov's Butterflies* 354)

If a platypus could mate with a duck this would exemplify the kind of convergence found in the smiles-boiler union. This, of course, is not possible. But what is possible is convergence between much simpler organisms or organisms with fairly close family ties. Single cell organisms, or simple sub-systems interacting within a complex system might easily couple in surprising ways.

I also call attention to the fact that, so far as it is concerned, smiles is conforming to its own linguistic laws. However, *we* can see that smiles' interpretation actually distorts its own natural linguistic laws. Because smiles is blind to the differences in boiler that are important to boiler, smiles has created new meaning and has a reproductive advantage over other communities. It can reproduce with more individuals because it is not restricted to its native community. Thus, smiles is now considered to have a higher fitness than it did before it discovered this additional way to reproduce.

The union of smiles and boiler and the creation of soiler may be considered an example of a natural mentalistic teleological event. The creation, soiler, is not *merely* subjective; it is a real viable individual. It will have real effects in the world. The union of smiles and boiler is a happy coincidence that, because it turns out to be useful, appears to have been caused or willed by someone above and beyond causality who can see into the unknowable future, caused the mutation in boiler and got him to the crossroads at exactly the right time for smiles' approach.

When an agent sees something as meaningful and uses it to increase its own reproductive fitness, a novel function has been discovered. All organisms interact and negotiate with their worlds by means of innumerable patterns that they have either inherited or have developed through self-organizing processes on their own. These patterns are "signs" because they are meaningful to the organism, that is, useful for self-creation and evolution. Signs may be materially embodied or may be purely behavioral. Some may be robust and flexible, some may be rigid, some may be quite vague. But signs are all around us; we grope our way; we make mistakes and grow, and if we're lucky, we live like artists do.

SIGNS IN PAINTING

In Rembrandt's *The Mill* (1645/48) we find a depiction of a country scene in the literal sense. If we don't notice the meaningful noise here, we say it's a painting of a windmill on a hill. But since we as humans often interpret vertical figures as ourselves, we impart an intelligence to the face of this mill: majestic, benignly lording over the toiling folk. The mill is also a cross, and as an energy-harnessing machine taking the place of Christ, it is profoundly significant on a level far below conscious experience. The mill is an accidental icon of a cross, that is, its blades could have described more of an X instead; mills are not inherently cross-like. We don't have to be consciously aware of these signs as signs; our bodies may unconsciously interpret them. Art has the power to convey meaning through resemblances that function as signs when noted by the self-organizing tendencies of our brains. And inanimate nature, I argue, can do the same with self-organizing tendencies of complex systems.

HENRY JAMES'S USE OF NOISE

The way nature transcends its own "laws" (directional tendencies) has been described above by the use of error and the misrecognition of patterns. Discovering accidental functions may be nothing but luck, but such luck seems to be an important part of artistic activity. One cannot *try* to discover accidental functions; we find them when we are looking for something else. In retrospect, we get the feeling that what we have found is what we *intended* all along. Teleology is always retrospective. That's why it's more favored by pondering artists and impractical philosophers than scientists who tend mostly to look ahead, not back.

Teleological luck is intentional because it isn't quite just luck. Some tend to be luckier than others for good reasons. To be so lucky requires an artistic temperament and a great deal of experience. Henry James' proximity to Peirce's theories through his brother William may help explain why he hit on a theory of artistry similar to Peirce's semiotics. Maybe, maybe not. But the fact of Henry's being an artist may very well explain (with or without Peirce) why Henry hit on them and William did not.

Henry James wrote a short story called "The Middle Years" (1893) late in his life. It's about a great author, not unlike himself, named Dencombe who befriends Hugh, a young doctor and his most astute critic. Dencombe confides in his young friend that he had worked according to an unconscious intention all his life, but that while he had worked, everything seemed always to come from trial and error:

> *What he saw so intensely today...was that only now, at the very last, had he come into possession. His development had been abnormally slow, almost grotesquely gradual. He had been hindered and retarded by experience, he had for long periods only groped his way.* (239)

Dencombe claims he can now see clearly the thing that he had "intended" all along, and he wants extra time to take advantage of his new insight and write now in a more teleological "new style." Dencombe wants a second chance to bring out his intentions more clearly. Hugh tells him that his work is wonderful just the way it is.

Because Dencombe describes his intention as if it had been there all along, and he had just failed to see it, I think of Dencombe as a Kantian, in contrast to Hugh, whom I think of as a pragmatist. It is no wonder that Dencombe did not recognize his intention as he was writing, because all teleological activity is only comprehended retrospectively. Dencombe feels that a predetermined whole is finally revealed in the end. He mourns the fact that, at the time, he could not comprehend its "law" and instead had to teach "himself by mistakes" (249). But Hugh wisely declares, "It's for your mistakes I admire you"(249). Hugh realizes that "the revelation of [Dencombe's] own slowness ... [made] all stupidity sacred" (251).

In James's short story, "The Figure in the Carpet" (1896), written at about the same time, a similar situation is investigated, but this time focused on the reader rather than the writer. In the story, a young critic (the unnamed narrator of the story) gets a rare chance to talk with his most admired novelist, Hugh Vereker, who reveals to him that his novels contain a hidden meaning, a "figure in the carpet" that no one yet has ever recognized. Vereker cannot explain what his intention is, point to exactly what causes it, or separate it from the medium in which it is expressed. Nevertheless, Vereker insists his life's work is governed by a "general intention"; he declares, "it chooses every word, it dots every i, it places every comma," and only he has recognized it (365).

This puts the young critic in a position similar to that of Biblical interpreter, described in Chapter Nine, who believes that there are hidden messages from God contained in the scriptures. The young critic is set on fire to discover the author's intention. He enlists the aid of his friend Corvick and his fiancé to try to decode the riddle, but their initial pedantic investigations prove fruitless.

New meaning can only come to the reader by chance. This is precisely what finally happens to Corvick: "when he wasn't thinking, [the words] fell, in all their superb intricacy, into one right combination" (381). Corvick's epiphany is a paradigm shift. Since one can only recognize what one already knows, novelty appears as noise to the observer who has no frame through which to understand it. As long as no paradigm exists to explain the noise, it will remain an anomaly. A new paradigm becomes possible when Corvick is exposed to new information, which has the effect of disrupting his usual patterns of perception. Corvick gets married. The newlyweds spend time in an unfamiliar place, India. Corvick is put in a position of interpreting what he does not understand in terms of what he does. After some time, however, the surroundings begin to enlarge his internal models of the world. Later, when he revisits the literary work, he begins to see it from a changed perspective.

Patterns he did not notice before suddenly cohere. Corvick had developed more plastic and structurally complex models of the world than he apparently had before marriage and before the trip to India. As boiler learned when smiles took advantage of him at the crossroads, intercourse with an "other" is one way of enlarging one's perspective. Corvick was then able (that is, he was lucky enough) to find an internal model that coincidentally fitted with the literary work in some significant way.

In these two stories Henry James contemplates the painful thought, shared by many excellent writers, that the true meaning of his work will never reach his audience. In the late 20th century the whine of such anxieties hit a crescendo and many abandoned all hope of communication, accepted the failure of intentionality and some even celebrated it. Although there is a melancholy note in these stories, James still knew, it seems to me, that communication is possible. It takes a special kind of reader, however, one who is as artistic as the writer, and who is willing to interact with the work over and over again until the reader actually becomes part of the work and his interpretations or responses begin to model the structure of the elusive wholeness that we call the meaning of the text.

CONCLUSION

In my arguments, I have not proceeded step by logical step. I've made long gently sloping arcs around the concepts of purpose and intentionality. In 1848 Edgar Allan Poe imagined—in a long prose poem-theory of the emergent origins of the universe—what someone hundreds of years in the future would say about the way we used to believe thinking got done. Poe thought our ancestors will think us absurd for having ever referred to the *path* of reason and *road* to knowledge. Reason, he argues, leaps from place to place. Teleological nature proceeds this way too, round about, indirectly—making puns in chemical reactions, neuronal networks, and social interactions, using metaphor and metonymy to find coincidental patterns—until finally a trail has been blazed through the chaos, and we have purpose.

Pragmatic teleology may have begun with Peirce and James over a hundred years ago, but it's still in its infancy. Art has had some success describing error-prone human interpreters, but this is about as far as this teleology has gotten. Biosemiotics might take it a lot further describing the artfulness of nature. Perhaps if art and science joined forces again, this teleology might really flower. The universe is perfused with signs, according to Peirce, and proto-interpretation is going on all around us in organic nature and to some extent in inanimate self-organizing systems too. When art-science can describe how nature reveals this, not just metaphorically but literally, then we will have truly teleological narratives.

Chapter 12

THE HAUNTED POSTMODERN

A living body is not a fixed thing but a flowing event, like a flame or whirlpool: the shape alone is stable, for the substance is a stream of energy going in at one end and out at the other. We are particular and temporarily identifiable wiggles in a stream that enters us in the form of light, heat, air, water, milk, bread, fruit beer, beef Stroganoff, caviar, and pâté de foie gras. *It goes out as gas and excrement—and also as semen, babies, talk, politics, commerce, war, poetry and music. And philosophy.*

–Alan Watts

Although Aristotle's teleology was long ago rejected as unscientific, and, more recently, the legitimacy of narrative itself (as a supposed godchild of determinism) has been put into question by the essential disconnectedness in the quantum world, it cannot be denied that out of this indeterminacy there has emerged the story of humankind, which, as physicist Paul Davies has gravely remarked, "seems almost contrived" (152).

Some believe that story-like organization cannot arise spontaneously in a world that is governed by chance and necessity alone. In the absence of a theory of self-organizing purpose such as I have outlined here, when intentional forms do emerge postmodernists are left to fear that some other force really is at work, guiding chance and putting the fix in on physical laws to suit supernatural purposes. Postmodern art and literature is haunted. It cannot be inhabited or ruled by gods or authors, because they are "officially" dead. And so it is left to try to bear the burden of its own absurd irony as best it can because *it does not believe in ghosts.* Although I've been critical of postmodernism throughout the work, I do believe that its fear of its own shadows gives it a great beauty. The ghostly suggestion of intentions in chance patterns can make any work of art or natural object intriguing.

Let's think back to William Paley's famous remark about how we should feel if we were to find a watch lying on the ground as opposed to a stone (he thought we would be right to infer a watch designer).[1] Now the postmodern question would be posed a bit differently. Let's suppose that trekking through a remote part of the jungle where no man has ever been before, we find a large upright stone. We know that tall stones don't usually stand upright because gravity tends to take them down. Why is this one standing? It seems improbable that it should have taken this position on its own. Its dimensions and position remind us of our human form, and so we have an inescapably uncanny feeling of being in the presence of another. It's not carved, fashioned or decorated in any way that would require a human hand or artist's touch, and yet this type of stone is not found in the area at all. What is the stone doing here? We wonder, with our bodies, if not with our conscious minds, does this stone *mean* something? But it can't! we sternly remind ourselves: it's just a fluke.

At its best postmodern art is like such a stone encountered in the wilderness.

I suppose I am, or have been, some kind of postmodern novelist and poet myself. (I hope to carry on with something else now as soon as I get this book done.) I am a product of it. I have been inspired by it. I am obsessed with its issues. Postmodernism's obscurities, and the fears and suspicions they just fail to cover, have made it the infectious, addictive trend it is. Dactyl Foundation has supported the research of those who appreciate postmodernism in this way, and this book should not be taken as an undervaluing of postmodernism's effectiveness and influence—not to mention eerie aesthetics, which others have rightly praised.

---

1. In *Natural Theology: Or Evidences of the Existence and Attributes of the Deity Collected from the Appearance of Nature* (1802), William Paley pointed out that we do not wonder how, say, a stone has come to be if we find it lying in a field, but if we were to find a watch instead, a watch with parts obviously organized to serve the purpose of keeping track of time, we would be bound to infer that it is a product of intention. By analogy then, a living organism is obviously organized to serve the purpose of its own survival; thus, so argues Paley, we are also bound to infer an intentional creator.

WHO LET THE GHOSTS IN?

Unjustly interned, the author returns to haunt the text.

Most critics agree that one of the defining characteristics of postmodern art and literature is its uncanniness. In earlier times, a miraculous event might have restored faith, reaffirmed beliefs or inspired a conversion. However strange, physically improbable, or otherworldly miracles were, the witnesses of ages past did not feel the uncanny like we do. The uncanny, as Freud noted, is felt more keenly by staunch nonbelievers, slapped with some insulting evidence that their reason and their understanding of the world may be seriously inadequate or simply wrong.

But the age in which we live now is not so different—in terms of its science—from the early 20th century. And so the question I have to ask is, Why did this feeling became more prevalent in the latter half of the 20th century? What makes the uncanny more possible now than sixty years ago? I believe postmodern experiments with formlessness may be the cause. Modernism, we may say, was overwhelmingly interested in form, not the lack thereof. If one begins with formlessness, it is more obvious, and startling, when patterns begin to emerge on their own. Postmodern novelists have consciously tried to write nonteleological "slice-of-life" narratives, preferring to string together random unconnected events and to include a lot of meaningless detail. They do so because they believe, as Paul Auster's narrator in "The Locked Room" says,

> *In the end, each life is no more than the sum of its contingent facts, a chronicle of chance intersections, of flukes, of random events that divulge nothing of their own purpose.* (257)

However, too often, they find, as I did under similar circumstances, that an uncanny sense of orderliness emerges from coincidental resemblances in the text. Because postmodern narratives tend to include a lot of random detail, there is a high probability that chance patterns will occur. Complexity and emergent order, argues Stuart Kauffman, require a high degree of random interaction before a system can self-catalyze. That's what we have, I think, in this kind of fiction. Thus these highly random narratives are often surprisingly reminiscent of the very teleological medieval knight's journey, which is basically a string of surprise meetings. Narrative theorist Mikhail Bahktin observes that it is precisely the randomness and chance in "adventure time" tales about chaotic life on the road that "provide an opening for the intrusion of nonhuman forces—fate, gods" (95).

Postmodern narratives, like analogical determined narratives, gesture toward an author outside of the text's narrative time who can mentally link things not physically connected. But, as Roland Barthes noted, these days that author is a dead one.

Postmodernists have tried to banish this ghost. They say, a pattern may point, but only at the infinite distance, tracing the curvature of the universe in a "straight" line, to return to the crown of its own particularity. They say, there is no other meaning there, only the literal material meaning here. Related to "allegories without ideas," noted by Renaissance scholar Angus Fletcher, postmodern stories can nevertheless still generate a vague spooky feeling that patterns are supposed to mean something. We live in a time when there is no official something. Ironically, we do respond to allegories without ideas as if there were ideas, probably because that's what animals with language do. We respond for the most part only as anesthetized automatons who have not paused to ask, What are those ideas exactly? What is it I'm buying? To ask the true meaning of anything might be symptomatic of a dangerous Platonism, and we've learned not to ask at all.

Angus Fetcher has argued, along with Emerson and Whitman, that our world is an aesthetic object of which we have knowledge (science) through mediated experience. His work, I may say, is the larger ecology in which this work plays its own idiosyncratic part. He has taught me that there is something fundamentally true about poetry, something not being realized often enough today. Knowledge of the world is not bits of information to be acquired step by step. It's a wild irreducible world of relations. Nature is a work of art.

I first met Fletcher in 1994 when I took a course on the "Literature of Nature" at the Graduate Center of the City University New York. Initially, the course sounded too pastoral for my tastes—in those days I thought I was interested in art not nature—and I almost passed it up, but the fates intervened in the form of Linda the all-knowing and benign department secretary, who said to me, "It's *Angus Fletcher*," and, without waiting for my response, wrote my name down on the roll. In this propitious way, I was introduced to the distinguished scholar, whose methods fitted perfectly with his subject and whose work has been, ever since, a steady spring of inspiration. In manner as well as appearance, with his vibrant white hair and beard, Fletcher might be described as a classic sage. A gifted professor, he tended to think aloud in class and to encourage students to join in as he wandered in his mental lake country.

Fletcher wrote the great book *Allegory: A Theory of a Symbolic Mode*. Ostensibly dedicated solely to literary subjects, what it does is explain how politics works. Allegorical stories can be used to seduce readers into following a static system of ideas that support a specific regime of power. It is not the function of allegories to invite questions; they draw their power from foregone conclusions. The ideas they represent are Platonic, *i.e.* eternal and universal. Like their counterparts the analogically determined teleologists, allegorists can make *anything* fit a preconceived notion; some can be really very good at this and extremely subtle. Edmund Spenser's *The Faerie Queene* (1596) is fashioned in a way such that every part reflects, like a fractal pattern, an idea about Queen Elizabeth's power. All the intricate, beautiful ritual and repetition—a stupendous amount of apparent diversity—is sameness, sameness all the way down. Thus allegory can have an anesthetizing effect, even as its beauty dazzles.

In the absence today in postmodern culture of Platonic ideas or Christian versions of them, stories read allegorically are more often used, no less manipulatively, to sell. These are "allegories without ideas." We buy the products advertised in sentimental literature and on television not really because we believe the uncertain promises but only because we have bought that product before, or one like it, and so has just about everybody else we know.

Postmoderns may distrust the allegorical kind of systematizing, but this hasn't stopped some system from self-organizing out of the anarchy they've fostered out of their fear of systems. Art looked on with a kind of fascinated abhorrence while commerce slipped in quietly and took the seats vacated by the church and government. Reducing themselves to ironic passivity, artists accepted their marginalized role because they couldn't think of an appealing counter-argument to commercialism that wasn't regressively Humanistic. Without the contributions of artistic individuals, the cheapest and easiest system is sure to emerge. And it has. We are inundated with cultural garbage. What emerges on its own is not necessarily good. Unfortunately, self-organization doesn't guarantee the best, only the most robust.

I think postmoderns have been unnecessarily concerned, and have, with the kind of zeal that infects too many revolutionaries, lumped teleological art in with propaganda. We need to be careful to make clear distinctions between self-organizing systems that are still adaptable and alive and self-organized systems whose directionality has become so entrenched that they have become incapable of responding to novelty or difference. Postmodern authors encountering their own self-organizing processes, simply don't quite know what to make of them, so they point to them ironically.

Despite his belief in the truly random nature of life, Paul Auster, for example, remains fascinated by the order that exists in art. Auster is, not incidentally perhaps, a former student of Angus Fletcher at Columbia, and so he probably learned from him, as I did, that allegory was once used to reinforce the Grand Narrative but is now used to sell soap, false happiness, and self-delusion. Like Kundera, Auster frequently uses involution, working himself into his own stories and calling attention to the artifice of fiction in general. In this way allegory can unbutton itself and let you know what's it's really up to. But we shouldn't suppose, as some have, that all meaningful art is a system of correspondences that point to an external, static idea. As Fletcher has made clear, allegory is a powerful symbolic mode among many.

THE HAUNTING

Martin Amis is a postmodern novelist who has explicitly grappled with the problems involved today with the idea of the writer as an omnipotent God of his creation. Amis was the subject of my first real research paper in college in 1993, and he was kind enough to grant me a phone interview. I had noticed that he was interested in the artfulness of life, and this encouraged me to believe teleology in art is a worthwhile topic to pursue. In *The Moronic Inferno*, he notes how coincidences make life feel "like a short story" that has an author above and beyond narrative time who can manipulate events. As an example, Amis recalls a visit to Chicago to meet novelist Saul Bellow at the Chicago Arts Club. It was a day, which, it seemed to Amis, the Fates had conspired to organize thematically. An artist materials store just outside Amis' hotel window bore a sign: "for the artist in everyone" (199). Back at his motel after a day of discussing the nature of art with Bellow, he notes, "the black, bent, bald shoeshiner who slicked my boots with his fingers (he had his name on his breast, in capitals) was called ART" (207). Amis's world seemed like Chesterton's that day.

Amis's awareness of life's fortuitous conjunctions is present in his mind when he sits down to make up a world. When he began to write his novel *Money: A Suicide Note*, he tried to include gratuitous descriptions and random details. He says he adopted "an inclusive, catch-if-catch-can attitude" resisting temptation to edit for coherence (Interview). Nevertheless, the novel's protagonist, John Self, begins to feel like he is in a novel with formal constraints when he, like his author, notices ominous coincidences. Self says,

> *something is waiting to happen to me. I can tell. Recently my life feels like a bloodcurdling joke. Recently my life has taken on* form. *Something is waiting. I am waiting. Soon, it will stop waiting—any day now.* (9)

Like Auster and Kundera (and many others of late), Amis includes himself as a character in his own novel—in order to show perhaps that the Author is too a fiction. When Self meets "Martin Amis" on the street he gets "the creeps" (61), but cannot quite figure out why. Throughout the novel, Self intuitively refers to people on the streets of New York as "bit players," "extras," and "actors," as if his life were controlled by an omniscient movie director. (Self doesn't read much, but he does watch a lot of movies, the medium of his Grand Consumption Narrative.)

Interestingly at the end of the novel, Self escapes the control of his author. Although Amis had planned to have Self commit suicide in the end, Self does not die. As (the real) Amis explained, in the end,

> *Self has escaped the novel. He's escaped control of the author figure, me. That's why that last section is in italics because it is, in a way, outside the novel. He really was meant to kill himself, but he screwed it up, as he screwed everything up. So, he's in a poorer but more controllable kind of existence. He feels that it's poorer because it is without form. It is more random, but that does suit him more or less. At least he's not being manipulated.* (see Alexander, "Martin Amis" 586-7)

Elsewhere Amis has claimed, "I have a god-like relationship [with] the world I've created. It is exactly analogous. There is creation and resolution, and it's all up to [me]" (McEwan). But clearly by his description of John Self's willfulness, we see that the author isn't like a tyrannical omniscient god at all, but like nature. Themes evolve and grow on their own and meanings come from within the world itself.

Amis's writing shows—his emergent Self shows—that he intuits something in art not quite captured by most theorists of deconstruction.

Hard-to-categorize-as-postmodern novelist Cormac McCarthy may speak to very different themes. There is an overwhelming fatalism in his work, but gods and demons don't work steadily manipulating people to suit their own purposes. Instead nature itself brings on its own destruction. In *Blood Meridian* (1985), the landscape is malevolent and regards the people trudging through it as pawns in a game ruled by its unalterable ends. We can say, with Harold Bloom, that the landscape is the most revealed character in the novel. We glimpse the landscape's recalcitrance and indifference in the poses it strikes, with its eerie lightening, meteor showers, ominous clouds, threatening mountains, and its hellish surfaces that stretch for miles without end. It is uncaring, harsh and without mercy. Just so the story ends unhappily in grotesque violence and injustice.

I met McCarthy in 2000 or so when we both were at the Santa Fe Institute (I doing research on teleology and insect mimicry, he writing fiction). During afternoon tea breaks, we tended to gravitate toward each other as the only two non-scientists present. Our conversations were mostly about postmodernism, for which he expressed disdain, disdain for the view of science as "just another narrative," and he noted how wrong some postmodernists had gotten the philosophical implications of quantum mechanics. It was my impression (memory is a bit dim) that McCarthy rejected both postmodern scientific relativism and cultural relativism, which allows an anything-goes permissiveness in art. McCarthy's stories certainly are not random. They are thematically coherent, and to say that there is a very strong authorial presence in his works is a colossal understatement. So McCarthy cannot be the typical sort of ironic postmodernist that I've been describing in this chapter. Nevertheless, the injustice driving his plots has aligned him with anti-Providential postmodern tropes.

As far as I know McCarthy still writes at SFI. I presume that he continues to enjoy the intellectual atmosphere there. The prevailing philosophy at SFI typically involves the belief that nature, as complex, is inherently unpredictable. Some have developed redefinitions of teleology similar to mine, allowing for a naturalized view of purpose as self-organizing and emergent. So then, even though the landscapes in McCarthy's worlds are uncanny, I cannot say that this feeling derives from the same sources of wonder and incredulity over self-organization that, say, Martin Amis's probably does. Instead I think it may have something to do with the belief that "end states" are final causes. Let me look into this and come back to McCarthy later.

THE FUTURE AS END

Sharing a taxi to the Santa Fe airport with a stranger last November, I mentioned my area of specialization was teleology. He suddenly started talking about "futurology" and how I ought to read this or that book. At the time I had no idea why he thought I might be interested in futurology. Only later did I realize that he had equated futurology with teleology. It's funny how you can develop such specialized interest in something that you no longer recognize your subject as it is for laymen. I suspect the source of the futurology-teleology connection is religious end-of-the-world tropes. As I spend most of my time thinking about teleology in the biological and semiotic worlds not in the religious world, this version of teleology has become unfamiliar to me, and so I was surprised by the suggestion. Imagine what a biologist would think if he were told he should read this or that book on futurology for his work.

Proponents of this teleology define purpose as "organized action to bring about future states." Therein lies the problem. This definition leaves out the all-important aspect of "aboutness" that is associated with intentionality. Purposeful actions (*i.e.* creatively self-maintaining responses) involve the representation (*i.e.* signs) of goal-objects not present. This is the more or less standard clarification offered by Franz Brentano (and T. L. Short after him) on the subject of intentionality. It's not *future states* so much as *signs of things not present* (and the entity's relation to them) that make actions purposeful.[2] In order to accept this futurology-teleology you have to believe that *telos* is linear, directed toward certain temporal end *states*. I've argued it is not.

"Future states" don't exist, therefore it doesn't make sense to talk about their causal effectiveness. This is just a little problem with the metaphoric language, but it can be confusing to general audiences, and it has even led some scientists to make arguments that the "final state" of a system is its "final cause." The final state of all systems is a point attractor, relative stillness, which for us means death. This is why I insist that a telic end should never be

2. All this is not missed so much as buried in the phrase "organized action." When you unpack "organization" it is necessary to bring in semiosis for the parts of an organized system are always constrained by signs of the whole.

associated with the "end state" as in "the last thing that happens." All complex systems tend to break down structure in their environments to maintain their own structures. In the process of doing so, they create more disorder in waste than they create order in themselves. At the end of a life, when purpose begins to fail, all the hard won physical structure starts to fall apart. (The meanings the entity created, however, may survive and continue to have effects, but how should these be measured against the loss of available energy to do work in the universe?) Life becomes compost for new life and so on until finally the sun's energy is degraded and gone and everything is dead. Some futurologists have reasoned that this is what life is *for*.

Obviously if you believe "representations of goal-objects not present" instead of "future states" constrain purposeful actions, the confusion about final cause being "in the future" disappears. But this just opens the door on another problem because normal science doesn't have a way to define "representations" properly—not in human thought any more than in a white blood cell. It needs to incorporate semiotics to do that. Following the lead of biosemioticians, especially Claus Emmeche, I have tried to marry semiotics with the complexity sciences. I have found Crutchfield's computation mechanics useful for this purpose, and I generalize semiotics to all complex systems.

Futurology comes out of a strain of teleology that does not argue, as I do, that semiotic purpose is physically embodied in structure and patterned activity. There is a tendency toward the denial of the flesh in such teleology. For example, as Schneider and Sagan note, German philosopher Arthur Schopenhauer (1788-1860) argued "in the manner of Buddhism, that the true spiritual goal [*i.e.* purpose] should be a release from the desires of acquisition and pleasure, power and success" (*Purpose of Life*). I argue, on the contrary, that purpose is always related to life and to growth, to imagination, and increasing complexity.

The "positive" take-home message of these religious futurology-teleologies is that humans—in believing that they ultimately serve grander purposes than just, say, pooping (*i.e.* dissolving gradients to return the universe to its original equilibrium or "oneness")—need to get toppled off their porcelain pedestals, need to be reminded of their heart-wrenching insignificance, need to stop imagining that they have selves that are in some small but extremely important sense decoupled from the energetic forces that control all things, and they need to accept the larger universal purpose of dissolving their egos into nothingness.

It is, perhaps, my deep-rooted Westernism that prevents me from signing on. Even though holism is ever so important to my theory, in general this work reflects my relative ignorance of Eastern literature, which, some can say, is where holism really has it roots. I quite admit I have an aversion to many things Eastern (due probably to its unfair association with New Age thinking). I haven't even been able to get myself to take yoga classes for fear that I might be encouraged to let go my ego and become one with the universe. I have not been able to accept Buddhism or any of its incarnations for I will not discount my ego. Making a self is no small task. *It's very hard to do well.*

Although we may only be typical of the kinds of intelligence that exists on earthlike or not-so-earthlike planets dotted all over the universe—or we may be even much less intelligent than our distant neighbors—I do not think that makes us any less special. Copernicus be damned. We, and our extraterrestrial siblings if they exist, can think things that are not. We can think relationships (analogies) into existence and change the course of history. An emergent self is the highest achievement. Its dissolution is an everyday tragedy.

That said, of course I do not advocate the greedy consuming type of Western selfhood: selves can only exist as part of a larger environment, which they co-create and co-maintain. Thus ecological sustainability is essential to truly purposeful action.

Human selves may be part of a larger ecology, which they must maintain as well as themselves, but I have a preference for thinking of them as the efficacious—and thus artistic—individuals that are in some sense apart from their environments. The notion of a hierarchical level (a whole) in science assumes the causal efficacy of such an entity. You may look at a part's actions as part of something else, or ignoring an entity's relation to something larger, you may look instead at how it's constraining its own parts. Wholes can be parts of a larger entity, but this does not mean they lose their sense of wholeness as they also act as part. Parts/wholes perform different roles in different contexts.

In preferring to see the self as part of a whole (rather than as a whole made up of parts), futurologists take *religious* teleology to its logical conclusion. Their theory of the purpose of life is consistent with the world's major religions,

Buddhism, Judaism, Christianity and Islam, which each conceive of life as ultimately the tool of the universal whole, which would not be diverse but utterly homogenous, with one purpose to return to divine equilibrium.

With his "death drive" idea, Sigmund Freud may have helped re-pave the way conceptually for this notion of an unconscious intention to become one with the universe—that is, to die. Freud made it possible for some to believe that whatever a man does accidentally is what he "really meant to do." For instance, a man whose drunk driving accidentally causes the death of his father, a passenger in the car, might be a man with an Oedipus complex who really wanted to kill his father so that he could "marry" his mother.

Now I have described motif-like mistakes as original purposeful behavior, and they can be purposeful so long as they reflect a general "theme" of the dynamical self, even as they alter it slightly. The question for the courts deciding man-slaughter charges in such a case would ask, Is this like something the accused will probably do again? Is this the sort of thing that he has done in the past? To find this out, character witnesses are called to relate narratives about the accused. As Alicia Juarerro has argued, it is only through the telling of stories that dynamical character can be revealed, and so this very subjective kind of evidence is needed in the courts. "Ample details also help jurors judge whether [the accused's] mental state unequivocally constrained the behavior or whether, to the contrary, noise might have compromised its flow into action" (*Dynamics* 238). The courts would have to decide: Is this act and example of his directionality? Or was it an accident?[3]

Most accidents are not purposeful, that is, they are not meaningfully incorporated such that they come to redefine character or bring out a previously latent potential. Most accidents just knock the wind out of us, and although we may regain ourselves afterwards, it is in spite of the accident not because of it. Life still has plenty of randomness in it. There are unintended side-effects of purposeful actions. Destruction is one of them. As I have shown

3. I said at the beginning of this book that I would leave the question of ethical responsibility to others more capable than I, and that I would focus on artistic responsibility. But this section does bring up questions regarding the *creative* criminal. What about the criminal who uses noise to determine his unlawful actions? Juarerro focuses on *directional* purpose in her analysis of intentionality as a complex system, so what she has to say about jurisprudence would apply, strictly, I would think, to sane perpetrators. If a criminal were to embrace dastardly mistakes, then that person may be an artist of sorts, but I think the more correct term might be "sociopath."

in an earlier chapter, Freud sometimes mistook true randomness for that which was intended. Freud with his "death drive" concept turns the failure of purposeful action into an active force with opposite ends.

"Purpose" in my sense always involves the creation and maintenance, the development and evolution of a self and/or of a response that defines that self. No one ever said purposes had to be eternal or universal. Indeed, as purposeful activity is synonymous with the activity of aliveness, purposes, like lives, naturally come and go.

Our purposes may turn out to be all for naught when we die. But one of the things that makes us purposeful in the first place is our ability to imagine and act on what might not be true. Intentionality refers to the ability to think about (and so to respond to) signs of things or qualities that may or may not exist, bringing ideas or representation into the causal matrix. Humans are very good at this. Their creative powers are exponentially greater than any other type of entity we know.

Throughout this book I have been playing down the significance of conscious decision-making and/or planning for purposeful action. Conscious deliberation undoubtedly involves the animal ability to think abstractly. We can say that animals ushered in a new kind of purposeful activity that didn't exist prior to their emergence on Earth (and wherever else they may be in the Cosmos). Conscious purposeful actions are selected/taken/chosen for the type of effect they are expected to have. Expectations involve abstractions. While most animals can use simple indexical and iconic signs, *we* appear to be fairly unique insofar as we can use conventional symbols and possess a language with a systematic grammar.[4] This means we are much better at both conscious and unconscious intentionality. We are different. There is no denying it: we write poems and build nuclear weapons.

I have argued that conscious intentionality tends to be directional (see Chapter Six), that is, when we consciously plan actions, we interact with our worlds according to well-established formal constraints that currently define our selves. When we interact with unconscious originality instead, we

4. See Terrence Deacon's *Symbolic Species*. There may be a few other primates raised under special conditions that have some rudimentary grasp of grammar. Perhaps whales or dolphins do; we don't know. I don't think humans are the only animals that can use grammar and conventional symbols, but we seem to be the only ones who do it extremely well at the moment.

tend to evolve new constraints. In this book I have stressed the importance of originality over directionality, not because I believe the former is more significant than the latter for defining purpose, but only because originality has been given less attention by others.

Although consciousness is not required for purposeful behavior, it does radically increase the effect of feedback—simple error correction—and, if combined with originality, it can also radically increase the effect of "reentry," to use Edelman and Tononi's term for massively parallel, simultaneous intercommunication among numerous regions of the human brain, which constantly *evolve new* definitions of rewards and errors. Conscious planning and consciously noting and editing useful chance patterns after the fact of their occurrence results in exponentially more complex purposeful behavior than that of which other animals are capable.

Some scientists predict that the universe is destined to end in thermal equilibrium. For fun let's imagine, because we can, that maybe something we do here in our short time in this starry sea will mean something to somebody somewhere. Maybe some agent will think of something useful to do with that side-effect entropy. Maybe travel beyond the speed of light is possible. Who knows? The fact that we can speculate on such things that we know to be untrue/impossible makes us wonderful intentional beings. We can act on what we imagine through signs. That makes us purposeful.

As a novelist I quite often contemplate this issue of the temporal end when I'm writing the end of a story. Does the last scene put everything into a new perspective? Endings can seem to function as a judgment about what the characters deserve. But teleology only pertains to development, evolution and agency: *why* entities emerge and act. A teleological nature cannot guarantee that whatever happens to someone is just. Sad things happen to teleological people. Good things do too. Being purposeful, actively creative or free doesn't protect one from making mistakes or having accidents or being at the wrong place at the wrong time. In fact, if one is not experienced, observant and sensitive, being more or less free to do as one pleases may greatly increase the probability of messing up. Tragedy, as defined by the example of the ancient Greeks, occurs when the hero, basically a good man, blazes a teleological path to his own destruction. We don't say Oedipus got what he deserved. We react with pity.[5] And even though we may say Oedipus acted purposefully, we do not say he purposefully sought to accomplish the death of his father and marriage to his mother. Things don't always work out the way we expect them to.

So then, to return to McCarthy who has been left waiting so patiently in the wings, I conjecture that the unhappy ending of his novel doesn't reflect a non-purposeful world. More likely it reflects the superior strength, cunning and/or luck of some people who happened to have destructive purposes. Who knows, maybe McCarthy didn't want the good and the kind to be murdered in his narratives, that's just the way their lives turned out.

Martin Amis's character John Self "got away" from him, and he could not give to him the ending that he had had in mind when he subtitled his novel. Self emerged as one ultimately not self-destructive. As the novel was being written, Self began to acquire a direction, which Amis came to realize in the process of creating him. This is what selves do, after all. Purposes emerge in the course of the life of a character, an organism, an entity, a system. They are not there from the beginning and they are not defined by the end.

5. This assumes that we more or less overlook Oedipus's crime of killing a stranger at the crossroads in the midst of his road rage. Did the ancients accept this as normal behavior or rationalize it as excusable hot-headedness common to all heroes?

EPILOGUE

A primary criticism of teleology is its supposed anthropomorphism. But it may be that there is a simple and profound truth to teleology that has been blighted by an association with a flawed notion of purposeful human behavior as (consciously) driven by an essential nature. Whether teleology is anthropomorphic or not, the way we understand human characteristics has changed since Aristotle named the fourth and most strange cause. Today, a new understanding of purposeful action as creative self-organization may be used to revise our understanding of telic phenomena and vice-versa. This reciprocal self-correcting tendency has ripened the fruits of teleology, which should not be left hanging on the vine.

Even so teleology has not and will not stop developing. Its end is still and always will be dynamic, something new, something larger, something greater, something more complex than anything we can measure or touch, something that guides and constrains but never pushes. In describing nature with metaphor and metonymy, art has explored many of her secrets, showing us her teleological ways. There are many things left to be explored, many things to be created, many more teleologies before us. Dactyl Foundation's discussions, lectures, research, and exhibitions are designed to keep secular teleology alive. We look for stories from all walks of life and want to hear your version of teleology. Together we might create a new era of artful, teleological emergence.

I will end, appropriately, not by spreading out into the lukewarm structureless space of artless nonbeing, but by offering some energy contained in poetry, while it's still available to us. This is an excerpt from "Blue Sonata" by John Ashbery, one of Dactyl Foundation's original board members. Often called an exemplary postmodernist, in this poem Ashbery sounds a little more like my kind of teleologist.

...There is a grain of curiosity
At the base of some new thing, that unrolls
Its question mark like a new wave on the shore.
In coming to give, to give up what we had,
We have, we understand, gained or been gained
By what was passing through, bright with the sheen
Of things recently forgotten and revived.
Each image fits into place, with the calm
Of not having too many, of having just enough,
We live in the sigh of our present.

If that was all there was to have
We could re-imagine the other half, deducing it
From the shape of what is seen, thus
Being inserted into its idea of how we
Ought to proceed. It would be tragic to fit
Into the space created by our not having arrived yet,
To utter the speech that belongs there,
For progress occurs through re-inventing
These words from a dim recollection of them,
In violating that space in such a way as
To leave it intact. Yet we do after all
Belong here, and have moved a considerable
Distance; our passing is a facade.
But our understanding of it is justified.

REFERENCES

Philosophy/Science

Adams, Frederick. "A Goal-State Theory of Function Attributions." *Canadian Journal of Philosophy* 9 (1979): 493-518. ISSN 0045-5091.

Alexander, Victoria N. "Nabokov, Teleology, and Insect Mimicry." *Nabokov Studies* 7 (2002/2003): 177- 213. ISSN 1080-1219.

---. *Narrative Telos: The Ordering Tendencies of Chance*. Diss. Graduate Center, City University New York, 2002. UMI 3047192.

---. "Neutral Evolution and Aesthetics: Vladimir Nabokov and Insect Mimicry." Working Papers Series 01-10-057. Santa Fe: Santa Fe Institute, 2001: 1-26.

---. "The Poetics of Purpose." *Biosemiotics* 2 (2009): 77-100. ISSN 1875-1342.

Aristotle. *Physics*. ca. 335 BCE. Trans. Robin Waterfield. New York: Oxford UP, 1996. ISBN 0199540284.

Augustine. *The City of God*. ca 410. Trans. Marcus Dods. New York: Modern Library, 1950. ISBN 0679600876.

Bacon, Francis. *Novum Organum*. 1620. Trans. Peter Urbach and John Gibson. Chicago: Open Court, 1994. ISBN 0812692454.

Baldwin, James M. "A New Factor in Evolution." *The American Naturalist* 30 (1896): 442-43. ISSN 00030147.

---. *Development and Evolution*. New York: Macmillian, 1902.

Bateson, Gregory. *Mind and Nature: A Necessary Unity*. New York: Dutton, 1979. ISBN 1572734345.

Bedau, Mark. "Against Mentalism in Teleology." *American Philosophical Quarterly* 27 (1990): 61-70. ISSN 0003-0481.

Bergson, Henri. *Creative Evolution*. 1907. Trans. Arthur Mitchell. Mineola, NY: Dover Publications, 1998.

Boden, Margaret. "Autonomy and Artificiality." *The Philosophy of Artificial Life*. Ed. Margaret Boden. Oxford: Oxford UP, 1996. 95-108. ISBN 0198751559.

---. "What is Creativity?" *The Dimensions of Creativity*. Ed. Margaret Boden. Cambridge: MIT Press, 1996. 75-117. ISBN 0262522195.

Bohr, Niels. *Philosophical Writings of Niels Bohr: Atomic Theory and the Description of Nature*. 1934. Woodbridge, CT: Ox Bow Press, 1987. ISBN 0918024501.

Boyd, Brian and Robert Pyle, eds. *Nabokov's Butterflies*. New York: Beacon Press, 2000. ISBN 0807085421.

Cassirer, Ernst. *Kant's Life and Thought*. 1918. Trans. James Haden. New Haven: Yale UP, 1983. ISBN 0300029829.

Clarke, Bruce and Mark B. N. Hansen, eds. *Emergence and Embodiment: New Essays in Second-Order Systems Theory*. Durham: Duke UP, 2009. ISBN 0822346001.

Crutchfield, James P. "Calculi of Emergence: Computation, Dynamics, and Induction." *Physica D* 75 (1994): 11-54. ISSN 0167-2789.

---."Is Anything Ever New? Considering Emergence." *Complexity: Metaphors, Models, and Reality*. Integrated Themes, Santa Fe Institute Studies in the Sciences of Complexity. Reading, MA: Addison-Wesley, 1994. 479-497. ISBN 0201626055.

---. "When Evolution is Revolution: Origins of Innovation." *Evolutionary Dynamics: Exploring the Interplay of Selection, Neutrality, Accident, and Function*. Eds. James P. Crutchfield and Peter Schuster. New York: Oxford UP, 2002. 101-134. ISBN 0195142659.

Crutchfield, James P., J. Doyne Farmer, Norman Packard, and Robert Shaw, "Chaos." *Scientific American* 255 (1986): 46-57. ISSN 0036-8733.

Darwin, Charles. *On the Origin of Species by Means of Natural Selection or the Preservation of Favored Races in the Struggle for Life*. 1872. Norwalk, CT: Easton Press, 1991.

Davies, Paul. "The Intelligibility of Nature," *Quantum Cosmology*. Ed. Robert John Russel *et. al*. Vatican: Vatican Observatory Foundation, 1996. ISBN 0268039763.

Deacon, Terrence. *The Symbolic Species: The Co-evolution of Language and the Brain*. New York: Norton, 1997. ISBN 0393038386.

Deacon, Terrence and Jeremy Sherman. "The Pattern which Connects Pleroma to Creatura: The Autocell Bridge from Physics to Life." *A Legacy for Living Systems: Gregory Bateson as Precursor to Biosemiotics*. Ed. Jesper Hoffmeyer. Vol. 2. Berlin: Springer, 2008. 59-76. ISBN 9048177030.

Dennett, Daniel C. *Freedom Evolves*. Viking: New York, 2003. ISBN 0670031860.

Derrida, Jacques. "Structure, Sign, and Play in the Discourse of the Human Sciences." 1966. Trans. Richard Macksey and Eugenio Donato. *The Critical Tradition: Classic Texts and Contemporary Trends*. Ed. David H. Richter. Boston: Bedford Book, 1989. 959-971. ISBN 0312101066.

Diaconis, Persi and Fredrick Mosteller, "Methods for Studying Coincidences." *Journal of American Statistical Association* 84 (1989): 853-861. ISSN 0162-1459.

Ducasse, C. J. "Explanation, Mechanism and Teleology." *The Journal of Philosophy* 22 (1925): 150-155. ISSN 0022-362X.

Edelman, Gerald and Giulio Tononi. *A Universe of Consciousness: How Matter Becomes Imagination*. New York: Basic Books, 2000. ISBN 0465013775.

Ehring, Douglas. "Goal-Directed Processes." *Southwest Philosophical Studies* 9 (1983): 39-47. ISSN 0885-9310.

Emmeche, Claus. "Closure, Function, Emergence, Semiosis and Life: The Same Idea? Reflections on the Concrete and the Abstract in Theoretical Biology." *Closure: Emergent Organizations and their Dynamics*. Eds. J. L. R. Chandler and G. Van de Vijver. New York: The New York Academy of Sciences, 2000.187-197. ISBN 1573312487.

Emerson, Ralph Waldo. *Emerson: Essays and Lectures.* Ed. Joel Porte. New York: Library of America, 1983. 941-968. ISBN 0940450151.

Favareau, Donald, ed. *Essential Readings in Biosemiotics: Anthology and Commentary.* Berlin: Springer, 2010. ISBN 1402096496.

Fontana, Walter and Leo Buss. "What Would be Conserved if the Tape were Played Twice?" *Proceedings of the National Academy of Sciences* USA 91 (1994): 757-761. ISSN 0027-8424.

Foerster, Heinz von. "On Self-Organizing Systems and their Environments." *Self-Organizing Systems.* Ed. Marshall C. Yovits and Scott Cameron. New York: Pergamon, 1960. 31-50.

Freeman, Walter J. *How Brains Make Up Their Minds.* New York: Co lumbia UP, 2001. ISBN 0231120087.

Freud, Sigmund. "The Uncanny." 1919. *The Complete Psychological Works of Sigmund Freud.* Trans. James Strachey. Vol. 17. London: Hogarth Press, 1974. 218-252.

---. "Determinism—Chance—And Superstitious Beliefs." 1914. *The Basic Writings of Freud.* Trans. and Ed. A. A. Brill. New York: Modern Library, 1995. 118-146. ISBN 067960166X.

Gardiner, Patrick, ed. *Theories of History.* New York: Free Press, 1959. 275-284. ISBN 0029112109.

Gell-Mann, Murray. *The Quark and the Jaguar: Adventures in the Simple and the Complex*. New York: Freeman and Company, 1994. ISBN 0805072535.

Godfrey-Smith, Peter. "Functions: Consensus without Unity." *Pacific Philosophical Quarterly* 74 (1993): 196-208. ISSN 0279-0750.

Goldstein, Jeffrey. "Emergence as a Construct: History and Issues." *Emergence: Complexity and Organization* 1 (1999): 49–72. ISSN 1521-3250.

---. *Emergence: Flirting with Paradox in Complex Systems: Understanding Emergence as Self-transcending Constructions*. Mansfield, MA: Emergent Publications, forthcoming.

---. "The Construction of Emergent Order, Or, How to Resist the Temptation of Hylozoism." *Nonlinear Dynamics, Psychology, and Life Sciences* 7 (2003): 295-314. ISSN 1090-0578

Goodwin, Brian. *How the Leopard Changed its Spots: The Evolution of Complexity.* New York: Scribner, 1994. ISBN 0025447106.

Goss, James. "The Poetics of Bipolar Disorder." *Pragmatics & Cognition* 14 (2006): 83-110. ISSN 0929-0907.

Gould, Stephen J. *The Structure of Evolutionary Theory*. Cambridge: Belknap Press, 2002.

---. "Eternal Metaphors of Paleontology." *Patterns of Evolution, as Illustrated by the Fossil Record.* Ed. A. Hallam. New York: Elsevier, 1977. 1-26. ISBN 0444414959.

---. *Wonderful Life: The Burgess Shale and the Nature of History.* New York: Norton, 1989. ISBN 0393027058.

Gould, Stephen J. and Richard C. Lewontin. "The Spandrels of San Marco and the Panglossian Paradigm: A Critique of the Adaptationist Programme." *Proceedings of the Royal Society B* 205 (1979) pp. 581-598. ISSN 0962-8452.

Hacking, Ian. *The Emergence of Probability: a Philosophical Study of Early Ideas about Probability, Induction and Statistical Inference.* Cambridge: Cambridge UP, 1975. ISBN 0521204607.

Harrit, Niels H., Jeffrey Farrer, Steven E. Jones, Kevin R. Ryan, Frank M. Legge, Daniel Farnsworth, Gregg Roberts, James R. Gourley and Bradley R. Larsen. "Active Thermitic Material Discovered in Dust from the 9/11 World Trade Center Catastrophe." *The Open Chemical Physics Journal* 2 (2009): 7-31. ISSN 1874-4125.

Hoffmeyer, Jesper. *Biosemiotics: An Examination into the Signs of Life and the Life of Signs.* Scranton: U of Scranton Press, 2008. ISBN 1589661699.

---. *Signs of Meaning in the Universe*. Bloomington, IN: Indiana UP, 1996.

James, William. "The Dilemma of Determinism." *The Will to Believe, and Other Essays In Popular Philosophy.* New York: Longmans, Green, and Co., 1897. 145-183. ISBN 0486202917.

Jones, Steven E. "Red Chips, Thermite." Boston 9/11 Conference. Faneuil Hall, Boston. 15 Dec. 2007. Available at <http://www.boston911truth.org/bostonconf.php>

Juarrero, Alicia. *Dynamics in Action: Intentionality as a Complex System.* Cambridge: MIT Press, 2002. ISBN 0262600471.

---. "Top-Down Causation and Autonomy in Complex Systems." *Downward Causation and the Neurobiology of Free Will.* Understanding Complex Systems. New York: Springer, 2009. 83-102. ISBN: 3642032044.

Kant, Immanuel. *Critique of Judgement.* 1790. Trans. J.H. Bernard. New York: Hafner Press, 1951. ISBN 0028475003.

---. "*Erklärung in Beziehung auf Fichtes Wissenschaftslehre.*" Public Declaration 6. 7 August 1799. Available at <http://www.korpora.org/Kant/aa12/370.html>.

---. "Idea of a Universal History from a Cosmopolitan Point of View." 1784. *Theories of History.* Ed. Patrick Gardiner. Glencoe: Free Press, 1959. ISBN 0029112109.

---. *Preface to Universal Natural History and Theory of Heaven: An Exploration of the Constitution and the Mechanical Origin of the Entire Structure of the Universe Based on Newtonian Principles.* 1755. Trans. Ian C. Johnston. Nanaimo, BC: Malaspina University, 1998. Available at <http://records.viu.ca/~johnstoi/kant/kant1.htm>.

Kauffman, Stuart. *The Origins of Order: Self-Organization and Selection in Evolution.* Oxford University Press, New York, 1993. ISBN 0195058119.

Keller, Evelyn Fox. *Century of the Gene.* Cambridge: Harvard UP, 2000. ISBN 0674008251.

---. "The Force of the Pacemaker Conception Theories of Aggregation in Cellular Slime Mold." *Reflections on Gender and Science.* New Haven: Yale UP, 1985. 150-151. ISBN 0300065957.

Kimura, Motoo. *The Neutral Theory of Molecular Evolution.* 1968. Cambridge: Cambridge UP, 1983. ISBN 0521231094.

Kuhn, Thomas. *The Structure of Scientific Revolutions.* Chicago: U of Chicago Press, 1962. ISBN 0226458083.

Kull, Kalevi, Terrence Deacon, Claus Emmeche, Jesper Hoffmeyer and Frederik Stjernfelt. "Theses on Biosemiotics: Prolegomena to a Theoretical Biology." *Biological Theory* 4 (2009): 167-173. ISSN 1555-5542.

Laplace, Pierre-Simon. *Philosophical Essay on Probabilities.* 1816. Trans. Andrew Dale. New York: Springer-Verlag, 1995. ISBN 0387943498.

Laughlin, Robert. *A Different Universe: Reinventing Physics from the Bottom Down.* New York: Basic Books, 2005. ISBN 046503828X.

Lenoir, Timothy. *The Strategy of Life: Teleology and Mechanics in Nineteenth Century German Biology.* Chicago: U of Chicago Press, 1989. ISBN 0226471837.

Linde, Andrei. "Eternally Existing Self-Reproducing Chaotic Inflationary Universe." *Physics Letters B.* 175 (1986): 395–400. ISSN 0370-2693.

Lovelock, James. *The Ages of Gaia: A Biography of Our Living Earth.* New York: Norton, 1995. ISBN 0393312339.

Lyotard, Jean-François. *The Postmodern Condition: A Report on Knowledge.* 1979. Trans. Geoff Bennington and Brian Massumi. Minneapolis: University of Minnesota Press, 1984. ISBN 0816611734.

Margulis, Lynn. *Origin of Eukaryotic Cells.* New Haven: Yale UP, 1971. ISBN 0300013531.

---. *Symbiotic Planet: A New View of Evolution.* New York: Basic Books, 1998. ISBN 0465072712.

Menand, Louis. *The Metaphysical Club: A Story of Ideas in America.* New York: Farrar, Straus & Giroux, 2001. ISBN 0374199639.

Millikan, Ruth Garrett. "Biosemantics." *The Journal of Philosophy* 86 (1989): 281-297. ISSN 0022-362X.

Nagel, Ernest. "Mechanistic Explanation and Organismic Biology." *The Structure of Science: Problems in the Logic of Scientific Explanation*. New York: Harcourt, Brace & World, 1961. 401-428.

Nissen, Lowell. *Teleological Language in the Life Sciences.* New York: Rowan & Littlefield, 1997. ISBN 0847686949.

Owen, Richard. *Lectures on the Comparative Anatomy and Physiology of the Invertebrate Animals.* London: Longman, Brown, Green, & Longmans, 1843. ASIN B0008CHFFI.

Oyama, Susan. *The Ontogeny of Information: Developmental Systems and Evolution.* Cambridge: Cambridge UP, 1986. ISBN 0521320984.

Paley, William. *Natural Theology: Or Evidences of the Existence and Attributes of the Deity Collected from the Appearance of Nature.* 1802. Oxford: Vincent, 1828.

Papineau, David. "Representation and Explanation." *Philosophy of Science* 51 (1984): 550-572. ISSN 0031-824.

Paracelsus. *Samtliche Werke.* 1589-91. Munich: O. W. Barth, 1922.

Peirce, Charles Sanders. *The Essential Peirce: Selected Philosophical Writings.* Eds. Nathan Houser and Christian Kloesel. Vol. 1. Indianapolis: Indiana UP, 1992. ISBN 0253207215.

Plato. *Plato's Cosmology: The Timaeus of Plato.* ca. 360 BCE. Trans. Francis Cornford. Indianapolis, IN: Hackett Publishing, 1997. ISBN 0872203867.

Plotnitsky, Arkady. "Two Conceptions of Chance, Teleology, and the Structure of Evolutionary Theory." 20th Annual Conference for the Society for Literature, Science and the Arts. Dactyl Foundation, New York. 10 Nov. 2006. Abstract <http://dactylfoundation.org/?p=43 >.

Reid, Robert G. B. *Biological Emergences: Evolution by Natural Experiment.* Cambridge: MIT Press, 2007. ISBN 0262182572.

Roqué, Alicia Juarrero. "Self-organization: Kant's Concept of Teleology and Modern Chemistry." *The Review of Metaphysics* 39 (1985): 107-135. ISSN 0034-6632.

Rosen, Robert. *Life Itself: A Comprehensive Inquiry into the Nature, Origin and Fabrication of Life.* New York: Columbia UP, 1991. ISBN 0231075642.

Russell, E. S. *Form and Function: A Contribution to the History of Animal Morphology.* 1916. Chicago: U of Chicago Press, 1982. ISBN 0226731731.

Sagan, Dorion. *Biospheres Metamorphosis of Planet Earth.* New York: McGraw-Hill, 1990. ISBN 0070544263.

Salthe, Stanley. *Development and Evolution: Complexity and Change in Biology.* Cambridge: MIT Press, 1993. ISBN 0262193353.

---. *Evolving Hierarchical Systems.* New York: Columbia UP, 1985. ISBN 0231060165.

Sanfillipo, L.C. and Hoffman, R. E. "Language Disorders in the Psychoses." *Concise Encyclopedia of Language Pathology.* Ed. Franco Fabbro. Oxford: Elsevier Science,

1999. 400-407. ISBN 0080431518

Schneider, Eric and Dorion Sagan. *Into the Cool: Energy Flow, Thermodynamics, and Life.* Chicago: U of Chicago Press, 2005. ISBN 0226739376.

---. *The Purpose of Life: Reconnecting Science and Religion.* Unpublished manuscript, 2010.

Short, T. L. *Peirce's Theory of Signs.* Cambridge: Cambridge UP, 2007. ISBN 0521843200.

---. "Measurement and Philosophy." *Cognitio: Revista de Filosofia* 9 (2008): 111-124. ISSN 1518-7187.

Taylor, Charles. *The Explanation of Behavior.* London: Routledge & Kegan Paul PLC, 1964. ISBN 0710036205.

Thom, René. "From Animal to Man: Thought and Language." 1975. *Essential Readings in Biosemiotics: Anthology and Commentary*. Ed. Donald Favareau. Berlin: Springer, 2010. 347-376. ISBN 1402096496.

Thomas. *The Summa Theologica of St. Thomas Aquinas.* 1265-74. Trans. Fathers of the English Dominican Province. London: Burns, Oates & Washbourne, 1920.

Thompson, D'Arcy Wentworth. *On Growth and Form.* Cambridge: Cambridge UP, 1917.

Thompson, Evan. "Life and Mind." *Emergence and Embodiment: New Essays in Second-Order Systems Theory.* Eds. Bruce Clarke and Mark B. N. Hansen. Durham: Duke UP, 2009. ISBN 0822346001.

Turing, Alan M. "The Chemical Basis of Morphogenesis." *Philosophical Transactions of the Royal Society B* 237 (1952): 37-72. ISSN 0962-8436.

Ulanowicz, Robert E. *A Third Window: Natural Life beyond Newton and Darwin.* West Conshohocken, PA: Templeton Press, 2009. ISBN 159947154X.

Watson, Andrew and James Lovelock. "Biological Homeostasis of the Global Environment: The Parable of Daisyworld." *Tellus B.* 35 (1983). 286–9. ISSN 0280-6509.

Watts, Alan W. *Does It Matter?: Essays on Man's Relation to Materiality*. 1970. New York: Novato, CA. New World Library, 2007. ISBN 1577315855.

Watts, Duncan. *Six Degrees: The Science of a Connected Age*. New York: Norton, 2004. ISBN 0393041425.

Weber, Bruce. "Design and its Discontents." *Synthese* 42 (2009): 837-856. ISSN 0039-7857.

Weber, Bruce and David Depew. "Natural Selection and Self-Organization: Dynamical Models as Clues to a New Evolutionary Synthesis." *Biology and Philosophy* 11 (1996): 33-65. ISSN 0169-3867.

Wegner, Daniel. *The Illusion of Conscious Will.* Cambridge: MIT Press, 2002. ISBN 0262731622.

Wheeler, Wendy. *The Whole Creature: Complexity, Biosemiotics, and the Evolution of Culture*. London: Lawrence & Wishart Ltd, 2006. ISBN 1905007302.

---. "A Failed Act of Eating: Appetite and Semiosymbiogenesis in a Biosemiotic Account of Aesthetic and Ethical Creativity," 20th Annual Conference for the Society for Literature, Science and the Arts. Dactyl Foundation, New York. 10 Nov. 2006. Abstract <http://dactylfoundation.org/?p=43 >.

---. "Creative Evolution: A Theory of Cultural Sustainability," *Communications, Politics and Culture* 42 (2009): 19-41. ISSN 0038-4526.

Wiener, Norbert. *The Human Use of Human Beings: Cybernetics and Society.* Boston: Houghton Mifflin, 1954. ISBN 0306803208.

Wimsatt, William. "The Ontology of Complex Systems," *Canadian Journal of Philosophy* 20 (1995): 564-590. ISSN 0045-5091.

---. "Teleology and the Logical Structure of Functional Statements." *Studies in History and Philosophy of Science* 3 (1972): 1-80. ISSN 1369-8486.

Woodfield, Andrew. *Teleology.* Cambridge: Cambridge UP, 1976. ISBN 0521211026.

Wright, Larry. "Explanation and Teleology." *Philosophy of Science* 39 (1972): 204-218. ISSN 0031-8248.

Narrative Theory/Poetics/Art Criticism

Alexander, Victoria N. "C. S. Peirce's Theory of Self-Organization and *The Crying of Lot 49*." *Pynchon Notes* 52 (2006): 23-39. ISSN 0278-1891.

---. "Martin Amis: Between the Influences of Bellow and Nabokov." *Antioch Review* 52 (1994):78-83. ISSN 0003-5769.

---. "Polonius and Poland, a Coincidence?" *English Language Notes* 36 (1999): 8-13. ISSN 0013-8282.

Alexander Victoria N. and Stanley Salthe. "Monstrous Fate: The Problem of Authorship and Evolution by Natural Selection." *Annals of Scholarship* 19 (1): 45-66. ISSN 0192-2858.

Amis, Martin. Telephone Interview. 8 May 1993.

---. *Koba the Dread: Laughter and the Twenty Million.* New York: Miramax, 2002. ISBN 0786868767.

---. *The Moronic Inferno: and Other Visits to America*. New York: Penguin, 1986. ISBN 0140127194.

---. "On Nabokov and Literary Greatness." Nabokov: A Centenary Celebration with PEN American Center, The New Yorker, and Vintage Books. Manhattan's Town Hall, New York. 15 April 1999.

Aristotle. *Poetics*. ca. 335 BCE. Trans. S. H. Butcher. New York: Hill and Wang, 1961. Available at <http://classics.mit.edu/Aristotle/poetics.html>

Auerbach, Erich. *Mimesis: The Representation of Reality in Western Literature.*1946. Trans. William R. Trask. Princeton: Princeton UP, 1969. ISBN 0691060789.

Bahktin, Mikhail. "Forms of Time and Chronotope in the Novel." *The Dialogic Imagination.* 1975. Trans. Kenneth Brostrom. Austin: U of Texas Press, 1981. 84-258. ISBN 029271534X.

Barthes, Roland. "The Death of the Author." *Image, Music, Text.* Trans. Stephen Heath. New York: Hill and Wang, 1968. 142-148. ISBN 0006861350.

Boyd, Brain. *Vladimir Nabokov: The American Years.* Princeton: Princeton UP, 1991. ISBN 0691024715.

Cleanth, Brooks. *The Well Wrought Urn*. New York: Reynal & Hitchcock, 1947. ISBN 0156957051.

Coleridge, Samuel. *Biographia Literaria: Biographical Sketches of my Literary Life & Opinions.*1817. Princeton, Princeton UP, 1985. ISBN 0691018618.

Culler, Jonathan. "Story and Discourse in the Analysis of Narrative." *The Pursuit of Signs: Semiotics, Literature, Deconstruction.* Ithaca: Cornell UP, 1981. 169-87. ISBN 0801487935.

Dimock, Wai Chee. "A Theory of Resonance." *PMLA* 10 (1997): 1046-59. ISSN 0030-8129.

Fletcher, Angus. *Allegory: A Theory of a Symbolic Mode*. Ithaca: Cornell UP, 1964. ISBN 0801492386.

---. "Allegory Without Ideas." *boundary* 2 33:1 (2006): 77-98. ISSN 0190-3659.

---. *Colors of the Mind: Conjectures on Thinking in Literature.* Cambridge: Harvard UP, 1991. ISBN 0674143124.

---. "Neil Grayson and 'This Living Hand'." Unpublished essay, 2009.

---. *A New Theory for American Poetry: Democracy, the Environment, and the Future of Imagination*. Cambridge: Harvard UP, 2006. ISBN 0674012011.

---. *The Prophetic Moment: An Essay on Spenser.* Chicago: U of Chicago Press, 1971. ISBN 0226253325.

---. *Time, Space and Motion in the Age of Shakespeare.* Cambridge: Harvard UP, 2007. ISBN 0674023080.

Forster, E.M. "Prophecy." *Aspects of the Novel.* 1927. New York: Harcourt Brace, 1955. 125-147. ISBN 0156091801.

Gilroy, James. *Don't Let Go.* Video Documentary. Dir. Neil Grayson. Ed. Chris Schwerin. Dactyl Foundation, 1999. Available at <http://dactylfoundation.org/?p=1288>

Grayson, Neil. Personal Interview. March 1997.

Jamison, K. R. *Touched with Fire: Manic-depressive Illness and the Artistic Temperament.* New York: Simon and Schuster, 1993. ISBN 0029160308.

Kermode, Frank. *The Genesis of Secrecy: On the Interpretation of Narrative*. Cambridge: Harvard UP, 1979. ISBN 0674345258.

Kugel, James L. "Early Interpretation: The Background of Late Forms of Biblical Exegesis." *Early Biblical Interpretation.* Auths. James Kugel and Rowan A. Greer. Philadelphia: Westminster John Knox Press, 1986. 13-106. ISBN 0664250130.

Kundera, Milan. "An Introduction to a Variation." *New York Times Book Review* 6 January, 1985, Late City Final Edition.

LaCapra, Dominick. "Trauma, Absence, Loss." *Writing History, Writing Trauma.* Baltimore: Johns Hopkins UP, 2000. 43-85. ISBN 0801864968

Lakoff, George and Mark Turner. *More than Cool Reason: A Field Guide to Poetic Metaphor.* Chicago, U of Chicago P, 1989. ISBN 0226468127.

Martin, Wallace. *Recent Theories of Narrative.* Ithaca: Cornell University Press, 1986. ISBN 0801417716.

McEwan, Ian. "Martin Amis with Ian McEwan." Writers Talk: Ideas of Our Time; Writers in Conversation Series. Northbrook, Ill: ICA Video, 1986.

Miller, J. Hillis. *Ariadne's Thread: Stories Lines.* New Haven: Yale UP, 1995. ISBN 0300063091.

Richardson, Joan. *A Natural History of Pragmatism: The Fact of Feeling from Jonathan Edwards to Gertrude Stein.* Cambridge Studies in American Literature and Culture Ser. 152. Cambridge: Cambridge UP, 2007. ISBN 0521837480.

---. *Wallace Stevens: The Early Years, 1879-1923.* Sag Harbor, NY: Beech Tree Books, 1986. ISBN 0688054013.

---. *Wallace Stevens: The Later Years, 1923-1955.* Sag Harbor, NY: Beech Tree Books, 1988. ISBN 068806860X.

Shklovsky, Viktor. *Theory of Prose.* 1925. Trans. Benjamin Sher. Elmwood Park, IL: Dalkey Archive Press, 1990. ISBN 0916583643.

Stravinsky, Igor. *Poetics of Music in the Form of Six Lessons.* 1947. The Charles Eliot Norton Lectures. Cambridge: Harvard UP, 1993. ISBN 0674678567.

Taylor, Richard. "Fractal Expressionism." Available at <http://www.uoregon.edu/~msiuo/taylor/art/fractal.pdf>

Todorov, Tzvetan. *Genres in Discourse.* Trans. Catherine Porter. New York: Cambridge UP, 1990. ISBN 0521349990.

Vincent, Steven. "Listening to Pop," *Antioch Review* 55 (1997): 96-105. ISSN 0003-5769.

Literary Works

Alexander, Victoria N. *Smoking Hopes.* Sag Harbor, NY: The Permanent Press, 1994. ISBN 1877946699.

Amis, Martin. *Money: A Suicide Note.* New York: Penguin, 1984. ISBN 0140088911.

Ashbery, John. "Blue Sonata." 1977. *Selected Poems.* New York: Penguin, 1986. 243-43. ISBN 0140585532.

Augustine. *Confessions.* ca. 398. Trans. R. S. Pine-Coffin. New York: Penguin Classics, 1961. ISBN 014044114X.

Auster, Paul. "The Locked Room." *The New York Trilogy.* New York, Penguin, 1990. 233-371. ISBN 0140131558.

Boethius, Anicus. *Consolation of Philosophy.* ca. 524. Trans. Victor Watts. New York: Penguin, 1969. ISBN 0140442081

Chaucer, Geoffrey. *The Canterbury Tales.* ca. 1380s. Trans. Nevill Coghill. New York: Penguin, 1977. ISBN 0140440224.

Chesterton, G. K., *The Innocence of Father Brown.* 1911. Seattle, WA: Createspace, 2009. ISBN 1449586619.

Diderot, Denis. *Jacques the Fatalist and His Master.* 1796. Trans. Michael Henry. New York: Penguin Books, 1986. ISBN 0140444726.

Hardy, Thomas. *The Pursuit of the Well-Beloved.* 1892. New York: Penguin, 1998. ISBN 0140435190.

James, Henry. *The Figure in the Carpet: and Other Stories.* 1888-1897. Ed. Frank Kermode. New York: Penguin, 1986. ISBN 0140432558.

Joyce, James. *Finnegans Wake.* 1939. New York: Penguin, 2000. ISBN 014118311X.

---. *Ulysses.* 1918-1920. Eds. Hans Walter Gabler with Wolfhard Steppe and Claus Melchior. New York: Vintage, 1986. ISBN 039455373X.

Karinthy, Frigyes. "Chain-Links." *Everything is the Other Way.* Trans. Adam Makkai and Eniko Janko. Budapest: Atheneum Press, 1929. Retrieved from <http://en.wikipedia.org/wiki/Six_degrees_of_separation>

Kundera, Milan. *Immortality.* Trans. Peter Kussi. New York: Grove, 1991. ISBN 0802111114.

---. *Jacques and His Master.* New York: Harper Perennial, 1985. ISBN 0060912227.

McCarthy, Cormac. 1985. *Blood Meridian: Or the Evening Redness in the West.* New York: Modern Library, 2001. ISBN 0679641041.

Nabokov, Vladimir. *Look at the Harlequins.* 1974. New York, Vintage, 1990. ISBN 0679727280.

---. *Lolita.* 1955. New York: Vintage, 1989. ISBN 0679723161.

---. *The Original of Laura.* 2009. New York: Knopf, 2008. ISBN 9780307271891.

---. "Ultima Thule." 1942. *The Stories of Vladimir Nabokov.* Ed. Dimitri Nabokov. New York: Knopf, 1995. 496-518. ISBN 0679729976.

Poe, Edgar Allan. *Eureka: A Prose Poem.* 1848. Kopenhagen: Green Integer, 1997. ISBN 1557133298.

Saint-Exupéry, Antoine de. *The Little Prince.* Trans. Katherine Woods. New York: Harcourt, Brace & World, 1943. ISBN 0152465030.

Shakespeare, William. *The Tragedy of Hamlet, Prince of Denmark.* ca. 1601. Ed. Harold Jenkins. New York: Routledge, 1995. ISBN 0415026830.

Sophocles. *Oedipus Rex.* ca. 429 BCE. Trans. Sir George Young. Mineola, NY: Dover Publications, 1991. ISBN 0486268772.

Stevens, Wallace. "The Idea of Order at Key West." 1936. *Collected Poetry and Prose.* Eds. Frank Kermode and Joan Richardson. New York: Library of America, 1997. ISBN 1883011450.

Vincent, Steven. "Switched Off in Basra." Editorial. *New York Times.* 31 July 2005. <http://www.nytimes.com/2005/07/31/opinion/31vincent.html>

---. *In the Red Zone: A Journey into the Soul of Iraq.* Dallas TX: Spence, 2004. ISBN 1890626570.

Voltaire, Francois-Marie Arouet. *Candide.* 1759. New York: Norton, 1991. ISBN 0393960587.

INDEX

A

B

C

D

E

F

G

H

I

J

K

L

M

N

O

P

Q

R

S

T

U

V

W